# LEGUME SEED NUTRACEUTICAL RESEARCH

Edited by **Jose C. Jimenez-Lopez**
and **Alfonso Clemente**

**Legume Seed Nutraceutical Research**

http://dx.doi.org/10.5772/intechopen.75158
Edited by Jose C. Jimenez-Lopez and Alfonso Clemente

## Contributors

Arindam Barman, Chinky M Marak, Rituparna Mitra Barman, Cheana S Sangma, Maria Bronze, Elsa Mecha, Maria Carlota Vaz Patto, Maria Eduardo Figueira, Alexandre Silva, Hanan Afifi, Ihsan Abu Alrub

## Notice

Statements and opinions expressed in the chapters are these of the individual contributors and not necessarily those of the editors or publisher. No responsibility is accepted for the accuracy of information contained in the published chapters. The publisher assumes no responsibility for any damage or injury to persons or property arising out of the use of any materials, instructions, methods or ideas contained in the book.

First published in London, United Kingdom, 2019 by IntechOpen
IntechOpen is the global imprint of INTECHOPEN LIMITED, registered in England and Wales, registration number: 11086078, The Shard, 25th floor, 32 London Bridge Street
London, SE19SG – United Kingdom
Printed in Croatia

British Library Cataloguing-in-Publication Data
A catalogue record for this book is available from the British Library

Additional hard copies can be obtained from orders@intechopen.com

Legume Seed Nutraceutical Research, Edited by Jose C. Jimenez-Lopez and Alfonso Clemente
p. cm.
Print ISBN 978-1-78985-397-1
Online ISBN 978-1-78985-398-8

# We are IntechOpen,
# the world's leading publisher of Open Access books
# Built by scientists, for scientists

**4,000+**
Open access books available

**116,000+**
International authors and editors

**120M+**
Downloads

**151**
Countries delivered to

Our authors are among the
**Top 1%**
most cited scientists

**12.2%**
Contributors from top 500 universities

Interested in publishing with us?
Contact book.department@intechopen.com

Numbers displayed above are based on latest data collected.
For more information visit www.intechopen.com

# Meet the editors

Jose C. Jimenez-Lopez has a BS in Biochemistry and Molecular Biology (1998), a BS in Biological Sciences (2001), and an MS in Agricultural Sciences (2004) from the University of Granada, Spain. He also has a PhD degree in Plant Cell Biology (2008) from the Spanish National Research Council (CSIC). Jose was a full-time postdoctoral research associate at Purdue University, USA (2008–2011), a Marie Curie Research Fellow (FP7-PEOPLE- 2011-IOF) (2012–2015) at the University of Western Australia and CSIC working in human health benefits of legume seed proteins, their allergy molecular aspects, and cross allergenicity, and is currently a senior research fellow (Ramon y Cajal research program—MINECO) working in the functionality, health benefits, and allergy implications of proteins from reproductive tissues (pollen and seeds) in crop species (mainly legumes) of agro-industrial interest.

Dr. Alfonso Clemente is a staff scientist at the Spanish National Research Council, working at the Estación Experimental del Zaidín (Granada, Spain). He has been working in legume seeds for the last 20 years, being involved in several national and international related projects. Alfonso Clemente joined different labs (Institute of Food Research, 1999–2000; John Innes Centre, 2000–2002; Sainsbury Laboratory, 2003–2004) in the UK to broaden his laboratory skills and scientific knowledge. Currently, he is the President of the Spanish Legume Association (Asociación Española de Leguminosas, www.leguminosas.es) having strong interaction with a relevant network of scientists and agricultural associations and agri-food companies in the field. He is author of more than 120 scientific manuscripts and an editorial board member of the *World Journal of Gastroenterology* and *Frontiers in Bioscience*, among others.

# Contents

# Preface

Legumes are grown worldwide and comprise the third largest family of flowering plants. They are a main source of nitrogen-rich edible seeds and constitute a major source of food for a significant part of the world's population, in addition to providing fodder, oil, fiber, and many other products.

Globally, grain legumes are one of the most relevant sources of plant proteins, providing also carbohydrates, minerals, vitamins, and other components with nutraceutical value and health benefit properties.

This book was conceived to provide key research knowledge on health-promoting aspects of seed components, and their nutritional and nutraceutical values and legume allergy features.

This book is intended to provide a series of peer-reviewed chapters that the editor believes will aid in increasing the quality of the research focus across the growing field of legume research. Overall, the objective of this project is to serve as a reference book and as an excellent resource for students, researchers, and scientists interested and working in different nutraceutical aspects of grain legumes, and particularly for the scientific community to encourage it to continue publishing its research findings on legumes, and to provide a basis for new research and the area of sustainable crop production.

**Jose C. Jimenez-Lopez, PhD**
Spanish National Research Council (CSIC)
Madrid, Spain

# Nutritional and Nutraceutical Properties of Legume

# Chapter 1

# Nutraceutical Properties of Legume Seeds and Their Impact on Human Health

Arindam Barman, Chinky M. Marak,
Rituparna Mitra Barman and Cheana S. Sangma

Additional information is available at the end of the chapter

http://dx.doi.org/10.5772/intechopen.78799

## Abstract

Legume seeds known to produce richer quality of proteins than cereals provide nutritious food for people around the world. Legume seeds contain around 20–40% protein. Apart from protein, it is also composed of carbohydrates, fiber, amino acids, micronutrients including several vitamins and minerals. Legume seeds can be considered a potent nutraceutical as it provides beneficial effects on human health as well as it helps in the prevention or treatment of certain diseases such as cardiovascular diseases, diabetes, digestive tract diseases, overweight, obesity, cancer, etc. Legume seeds also contain anti-nutritional compounds which may be toxic when consumed raw, but when processed and treated may play a positive role on human health. There are many more underutilized food legume seeds that may be a potential source of nutraceutical food. The main aim of this chapter is to describe the nutraceutical properties of legume seeds and their impact on human health.

**Keywords:** legume, nutraceutical, anti-nutritional, underutilized legume, human health

## 1. Introduction

Nutraceutical can be defined as a food or part of a food that provide medical or health benefits, including the prevention or treatment of a disease [1]. It can be isolated nutrients, dietary supplements, specific diets, designer foods, herbal products, processed foods or processed beverages. Several nutraceuticals found in legumes are listed among the top 200 list of the American Nutraceutical Association [2]. Legumes are considered to produce substantial amount of proteins than cereal grains. Not only proteins but legumes also supply adequate amount of energy, carbohydrates, minerals, vitamins, and dietary fiber with low fat production [3]. The major

storage proteins of legume seeds are oligomeric globulins which are of 7S and 11S protein fractions [4]. Legumes are well known for the presence of different bioactive compounds such as saponins, tannins, flavonoids, isoflavones, lectins, phytic acid etc. which is important for its nutraceutical property [5, 6]. Highly pigmented and dark colored legume seeds have higher level of phenolic and flavonoid content which helps in its antioxidant activity [5, 7–9]. Legume seeds contain enzyme inhibitors like $\alpha$-amylase, $\alpha$-glucosidase and $\gamma$-aminobutyric acid (GABA) for which it can be used as a nutraceutical molecule. Green legume seeds are also a good source of nutraceuticals [10]. Legume seeds are normally consumed after processing there by increasing the bioavailability of nutrients by inactivating trypsin, growth inhibitors and hemagglutinins [11]. Different species of legumes are involved in the treatment of various diseases like coronary heart diseases, cardiovascular diseases, cancer, diabetes, etc. [5, 7–9]. Legume seeds also contain resistant proteins and carbohydrates that play an effective role in human health [12]. The importance of legumes in human diet is expected to increase in the near future in order to meet the demand for protein and other nutrients in the increasing world population and also to reduce the risk related to animal food source consumption. Molecules present in legume seeds that are considered toxic or unhealthy may also provide positive effects on human health in the prevention and treatment of certain diseases if consumed in a limited scale and proper way, and hence play an excellent role in the nutraceutical and antioxidant property of seed legume [4].

## 2. Different species of legumes possessing a potential nutraceutical property

### 2.1. Black soybean (*Glycine max* L.)

It is a variety of soybean composed of black seed coat that has been extensively used as a tonic food and material in oriental medicine for many years. The traditional Chinese medicine theory believes that black soybean is helpful in treatment of diabetes, hypertension, anti-aging, cosmetology, blood circulation, etc. due to its active peptide compounds [13].

### 2.2. Pegion pea (*Cajanus cajan* L. Millspaugh)

This legume food is an excellent source of protein, starch, calcium, manganese, crude fiber, fat, trace elements and minerals. Pigeon pea seeds are composed of 85% cotyledons, 14% seed coat, about 1% embryo and a variety of dietary nutrients. The embryo contains majority of the seed proteins whereas the cotyledons constitute majority of the carbohydrates. It has both the nutritional and medicinal property. Scorched seeds can relieve headache and vertigo when added to coffee while fresh seeds help urinary incontinence in males. On the other hand, immature seeds are used in the treatment of kidney ailments. Pigeon pea seed husks possess an effective anti-oxidant and anti-hyperglycemic activity which may be a potential organic resource for the development of nutraceutical for hyperglycemic individuals [14].

### 2.3. Mung bean (*Vigna radiata*)

It is rich in proteins, carbohydrates, amino acids and vitamins. It contains different bioactive compound which help in lowering the risk of various diseases [5].

## 2.4. Cowpea (*Vigna unguiculata*)

It is an important leguminous food rich in protein, carbohydrates, minerals, and water soluble vitamins like thiamine, riboflavin and niacin [5].

## 2.5. Rice bean (*Vigna umbellate*)

It is known as climbing mountain bean, mambi bean and oriental bean and is native to Southeast Asia. It has high nutritive value with rich source of protein and essential amino acids such as lysine, tryptophan and methionine. It also contains a number of bioactive compounds such as phytate, $\alpha$-galactosides and trypsin inhibitors which can act as anti-oxidant, anti-cancer and anti-diabetic agents [14].

## 2.6. Black gram (*Vigna mungo*)

It is an important legume crop having high nutritional value used as a diet during fever, cooling astringent, poultice for abscesses, affection for cough and liver and also recommended for treating diabetes [8].

## 2.7. Lentil (*Lens culinaris*)

It is a rich source of proteins, vitamins, minerals, dietary fibers, folic acid and carbohydrates, mostly the resistant starches. It also contains different bioactive compounds such as lectins, enzyme inhibitors, phytates, oligosaccharides, and phenolic compounds. Lentil seed is composed mostly of carbohydrates. Lentil seeds play an important role in the prevention and treatment of various diseases. Due to the high content of dietary fiber and low glycemic response of lentil seeds, it is highly recommended for patients suffering from cardiovascular diseases and diabetes. Several bioactive compounds present in lentil seeds such as phytic acids, lectins, defensins, saponins, etc. show anti-carcinogenic, anti-mutagenic, anti-oxidative and anti-hyperglycemic activities [15].

## 2.8. Chick peas (*Cicer arietinum*)

The demand for chick pea is high due to its nutritional value. In the semi-arid tropics, chickpea is an important component of the diets of those individuals who cannot afford animal proteins. Chickpea is cholesterol free and is a good source of carbohydrates, protein, dietary fiber (DF), vitamins and minerals [16, 17]. Chickpea consumption has been reported to reduce the risk of chronic diseases and optimize health. Chickpea seed oil contains different sterols, tocopherols and tocotrienols [18]. These phytosterols have been reported to exhibit anti-ulcerative, anti-bacterial, anti-fungal, antitumor and anti-inflammatory properties coupled with a lowering effect on cholesterol levels [19].

## 2.9. Lupins (*Lupinus* sp)

Lupin seeds contain many bioactive components. The protein, which may correspond to 35–40% of the dry weight, is mostly composed of albumins and globulins in a ratio of 1:9 [20]. Different potential health benefits of lupin have been investigated, particularly in the area of dyslipidemia, hyperglycemia, and hypertension prevention [21].

## 2.10. Peanut (*Arachis hypogaea*)

Peanuts are a rich source of omega-3, fiber, vitamin E, antioxidants and "good" fats. Consumption of peanuts has been associated with a number of health benefits, particularly for the heart and to reduce the risk of blood clots, lower cholesterol and reduced the risk of arrhythmia. Now, a recent study supports that the consuming peanuts may protect against death from numerous diseases, including cancer, heart disease and diabetes [22] (**Figure 1**).

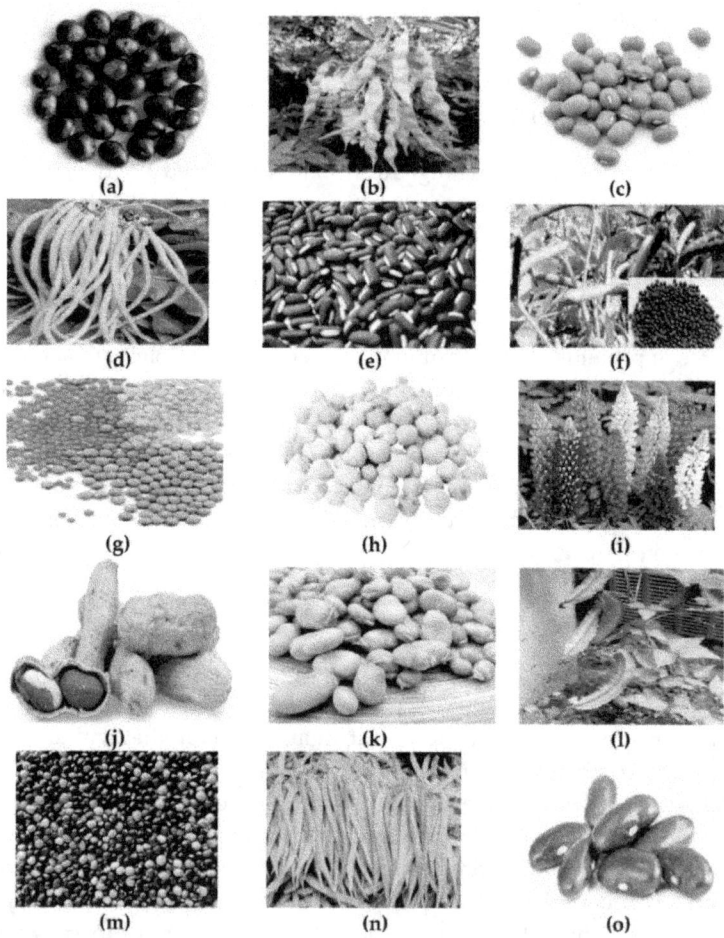

**Figure 1.** Species of legumes possessing potential nutraceutical property [a. Black soybean (*Glycine max* L.); b. Pegion pea (*Cajanus cajan* L. Millspaugh); c. Mung bean (*Vigna radiata*); d. Cowpea (*Vigna unguiculata*); e. Rice bean (*Vigna umbellate*); f. Black gram (*Vigna mungo*); g. Lentil (*Lens culinaris*); h. Chick peas (*Cicer arietinum*); i. Lupins (*Lupinus sp*); j. Peanut (*Arachis hypogaea*); vetch (*Vicia faba*); l. winged bean (*Psophocarpus tetragonolobus*); m. Pigeon pea (*Cajanus cajan*); n. Cluster bean (*Cyamopsis tetragonoloba*); o. Kidney bean (*Phaseolus sp*)].

## 2.11. Vetch (*Vicia fava*)

*Vicia* is a genus of about 140 species of flowering plants that are part of the legume family (Fabaceae), and which are commonly known as vetches. The vetch or fava bean is an important food crop, and several other species of vetch are cultivated as fodder and cover crops and as green manure. It is a protein-rich legume seed and also at the same time toxic to humans if consumed in quantity. The seed composed of tannins, vicine and convicine. Beside the positive impact of tannin-free varieties, the development of faba bean cultivars with very low levels of vicine and convicine would represent a real advantage in terms of nutritional performance in poultry diets and of food safety to humans [23].

## 2.12. Winged bean (*Psophocarpus tetragonolobus*)

It is a source of important minerals, such as iron, manganese, copper, calcium, phosphorus, magnesium. It also contains an abundance of Vitamin A, which is a powerful antioxidant that prevents DNA damage [24].

## 2.13. Pigeon pea (*Cajanus cajan*)

Pigeon peas are rich in proteins, minerals, vitamins and lipids. It is an excellent source of magnesium, phosphorus, calcium and potassium. It provides an adequate amount of iron and selenium. Pigeon peas contain dietary fiber, potassium and low cholesterol which help to maintain the healthy heart. Potassium lowers the strain on heart by reducing the blood pressure. Dietary fiber maintains cholesterol balance and prevents atherosclerosis [25].

## 2.14. Cluster bean (*Cyamopsis tetragonoloba*)

Cluster beans are rich source of soluble fiber content and are known for their cholesterol lowering effect. It is also beneficial for blood circulation, resolve anemia, make bone strong and stimulates bowel movement [26].

## 2.15. Kidney bean (*Phaseolus* sp)

Phaseolus is the most important food legume for human consumption in the world. Its seeds consist mainly of carbohydrates and are a good source of nitrogen and protein. It also contains calcium, magnesium, potassium, phosphorus, copper, iron, zinc, manganese and sulfur. This legume is rich in bioactive components such as enzyme inhibitors, lectins, phytates, oligosaccharides and phenolics, which exhibit metabolic roles in humans and animals. Among the observed biological activities are the antioxidant capacity, the reduction of cholesterol and reduction of low-density lipoproteins, thus Phaseolus has a protective effect against cardiovascular diseases. Also it has shown favorable effects against cancer because of the antimutagenic and antiproliferative properties of their phenolics, lectins and protease inhibitors. Additionally, it has showed effects on obesity and diabetes due to its content of resistant starch and $\alpha$-amylase inhibitor [7].

# 3. Nutritional and anti-nutritional components found in legume seeds

There are various nutritional and anti-nutritional components present in legume seeds which plays a significant role as a nutraceutical property of legume seeds.

## 3.1. Phenolic compounds

In legume seeds, phenolic compounds are found as antinutritional compounds, but it also can act as antioxidants because of its ability to chelate metal ions, inhibit lipid peroxidation and scavenge free radicals. Phenolic compounds found in legume seed are mainly tannins, phenolic acids, anthocyanins and flavonoids Polyphenolic compounds such as flavonol glycoside, anthocyanins and condensed tannins provide color to the seeds of legumes. The seeds which are dark colored and highly pigmented have high phenolic content such as red kidney beans (*Phaseolus vulgaris*) and black gram (*Vigna mungo*). The phenolic content of the legume seeds is highly associated with its anti-oxidant activity [5, 15]. Phenolic compounds in legume seeds show anti-bacterial, anti-viral, anti-inflammatory and anti-allergenic activities. They are also known to lower the risk of cancer, heart diseases and diabetes [8]. It has been reported on the analysis of total phenolics against antioxidant activities that fermented legume seeds exhibit more antioxidant potential [27]. The balance between antinutrient and biological antioxidant effects present in legume seeds will help in nutrient utilization improvement thereby providing potential nutraceuticals for human health [8].

## 3.2. Tannins

Tannins help in removal of toxins from the intestine because of its ability to bind the proteins. Tannins also help in maintaining hygiene of the mouth by inhibiting bacterial growth that cause tooth decay [5, 27].

## 3.3. Flavonoids

Flavonoids contained in legume seeds act as an antioxidant and prevent many diseases such as cancer, cataract, arteriosclerosis, autoimmune diseases, inflammation and aging [4–6].

## 3.4. Isoflavones

Isoflavones in legume seeds act as an antioxidant and helps in lowering the risk of many diseases such as osteoporosis, cardiovascular diseases and cancer. It is also used for the treatment of menopause symptoms [5]. Isoflavones daidzein and genistein are natural phytoestrogens that are able to inhibit LDL oxidation thereby reducing the risk of atherosclerosis [28]. Genistein and Daidzein acts as an anti-cancer agent. Genistein helps in the inhibition of platelet aggregation, leukotriene production, DNA topoisomerase II, angiogenesis, reduction of bioavailability of sex hormones, induction of apoptosis, and differentiation in cancer cells. Daidzein induces differentiation in B16 melanoma, and HL-60 human leukemia cells. Daidzein also helps in the inhibition of enzyme ALDH-I, an NAD dependent aldehyde dehydrogenase that catalyzes the oxidation of acetaldehyde, the primary product of alcohol metabolism which further helps in the treatment of alcoholism. There is epidemiological evidence regarding high legume seed diets to reduced risk of cancer [29].

### 3.5. Phytic acid

Phytic acid is present in legume seeds as antinutrients. It is stored in legume seeds in the form of phosphate in the endosperm [15] Phytic acid act as anti-HIV agent by inhibiting the transcription of the viral genome. Phytic acid prevents the formation of kidney stones. It also helps in the prevention of cavities, plaque and tartar in the teeth by reducing solubilities of calcium, fluoride and phosphate and protects it from demineralization. Phytic acid also helps in reducing risk of heart diseases and diabetes mellitus [7]. It also shows antioxidant and anticarcinogenic properties [30].

### 3.6. Saponins

Saponins are secondary plant metabolites which can be either steroidal or triterpenoid. Commonly found saponins in legume seeds are triterpenoids. Saponins may play a significant role as anti-cancer agent by reducing the formation of carcinogenic substances in colon. They may also lower the risk of heart diseases. Saponins may also show immune stimulant effects by inducing the formation of cytokinins such as interleukins and interferons. Saponins may also provide several other beneficial effects such as anti-inflammatory, anti-fungal, anti-parasitic, hypercholesterolaemia, hypoglycemia, immunomodulatory, etc. [15, 30].

### 3.7. Oligosaccharides

Legume seeds constitute indigestible substances especially flatulence induced oligosaccharides, e.g. raffinose, stachyose and verbascose. These oligosaccharides can be used as a prebiotic agent by promoting the growth of bifido bacteria. They may show anti-carcinogenic, anti-diabetic and anti-cardiovascular effect and also a higher rate of mineral absorption that are beneficial to human health [8].

### 3.8. Enzyme inhibitors

Legume seeds contain enzyme inhibitors which are involved in the regulation of endogenous proteases, amylases, lipases, glycosidases and phosphatises enzyme. They provide defense mechanism against seed eating insects and microorganisms. Enzyme inhibitors of legume seeds belong to Kunitz (20–24 kDa) and Bowman-Birk (8 kDa) family [5].

#### 3.8.1. Protease inhibitor

Seeds of legumes such as pea, chickpea and mung bean protein hydrolysates have angiotensin converting enzyme (ACE) inhibitory activity which helps in improving cardiovascular health. Inverse correlation between pulse consumption and risks of colon cancer, prostrate cancer, gastric cancer and pancreatic cancer shows that it helps in the prevention of these diseases [5, 28].

### 3.8.2. α-Amylase inhibitor

α-Amylase protein inhibitor present in raw legume seeds help in the prevention and therapy of obesity and diabetes [5, 28].

### 3.8.3. α-Glucosidase inhibitor

Legume seeds contain α-glucose inhibitor which helps in the treatment of diabetes. Postprandial hyperglycemia, the earliest metabolic abnormality of type 2 diabetes mellitus can be suppressed by inhibition of α-glucosidases delaying the digestion and absorption of carbohydrates [5, 28].

### 3.8.4. γ-Aminobutyric acid

γ-Aminobutyric acid (GABA), a non-protein amino acid, acts as a depressive neurotransmitter in the sympathetic nervous system. GABA is effective for the treatment of hypertension, depression, sleeplessness, autonomic disorders, and chronic alcohol related symptoms and for stimulation of immune cells [5, 28].

## 3.9. Polysaccharides

The presence of carbohydrate in grain legume seeds in amounts up to 60%, which include oligosaccharides, such as alpha-galactosides, and complex molecules, such as starches and fibers [31]. Starch is a storage polysaccharide made up of amylose and amylopectin in the ratio roughly of 1:3 but in wrinkled pea the ratio reach 3:1 [32]. As far as starch concerns, in lupin and soybean are found the lowest amounts (about 1–2%) whereas in pea and fababean it accounts for about 50% of the dry seed weight. From the nutritional viewpoint this polysaccharide can be classified, according to its hydrolysis degradation in animal model systems, as rapid digestion starch (RDS), slow digestion starch (SDS) and resistant starch. This later is not hydrolysed by human amylases but it can be fermented by the microorganisms present in the colon as if it is fiber [33].

## 3.10. Lipids

Lipids are generally present in legume seeds such as pea, common bean, lentil at low amounts (1–3%) with approximately 40–50% neutral lipids, 25–35% phospholipids and 9% glycolipids. Phospholipids have been shown to possess potent effects on serum lipoproteins and the phospholipids in pulses could contribute to the lipoprotein effects [34]. The effects on the cardiovascular system in preventing diseases are seemingly due to their capacity to lower the blood cholesterol levels by inhibiting cholesterol absorption [35].

## 3.11. L-DOPA

Seeds of wild legume *Mucuna pruriens* and seeds of *C. cathartica* are excellent natural source of the neurotransmitter L-DOPA. Consumption of food possessing L-DOPA is toxic only to those individuals having deficiency of glucose-6-phosphate dehydrogenase (G-6-PD) in their erythrocytes. Although L-DOPA is neurotoxic, it is highly effective to treat the Parkinson's disease [36–39]. L-DOPA is responsible for the release of human growth hormone (HGH) from pituitary gland and it can be effectively used in the treatment of restless leg syndrome (http://www.ihealthdirectory.com/l-dopa/). Administration of L-DOPA for a long duration to the patients suffering from Parkinson's disease, there were no adverse effects to prove its toxicity.

### 3.12. L-Canavanine

L-canavanine is a suitable inhibitor of cancer-inducing nitric oxide [40]. It can prevent the anti-tumor activity of Walker carcinoma, human melanoma, pancreatic cancer [41, 42], plant sources containing L-canavanine in therapy of colon cancer through diet management. L-canavanine is highly suitable for pancreatic cancer studies due to lack of considerable amount of arginase in the pancreas [41].

### 3.13. Lectins

The lectin con C of *C. cathartica* has a number of applications such as a blood grouping substance [43], immunomodulator [44] and tissue marker [45]. Con A induces immune cell responses in liver, which leads to destruction of tumor cell, after binding to mannose moiety, con A exerts both autophagic and anti-hepatomic (immunomodulation) properties. Adeno virus micro-beads bound to con A is an effective agent in delivering therapeutic transgenes for inflammatory bowel disease [46]. It has also been found that con A stimulates mitosis and inhibits the lymphocyte cap formation and patch formation due to anti-immunoglobulin. In isolated fat cells, con A shows similar activity to that of insulin [47].

### 3.14. Proteins

Mature seeds of pea, faba bean and beans contain 18–20% protein and lupin and soybean contains 35–45% protein. Most of the proteins found in legume seeds are storage proteins which are of 7S and 11S globulins based on their sedimentation coefficient [4]. Some proteins in legume seeds show antifungal and anti-viral activity and hence act as anti-HIV and anti-diabetic agent [5]. These proteins contain several essential amino acids which are beneficial to human health.

The proteins present in *Vigna species*, show antifungal and antiviral activity. Ground bean lectin inhibits the hemagglutinating activity by polygalacturonic acid but not galacturonic acid and simple monosaccharides [31]. It decreases the viability of hepatoma (HepG2), leukemia (L1210) and leukemia (M1) cell and also elected a mitogenic response from mouse splenocytes. Due to presence of all these properties, these proteins act as an excellent drug for the treatment of AIDS patients with no adverse effects as compare to synthetic drugs [44].

### 3.15. Minerals and vitamins

Minerals act as a cofactor for many enzymatic reaction e.g. copper, zinc, and magnesium and manganese. Vitamin E and C are known to play a role as an antioxidant and inhibiting the oxidation of vitamin A in the gastric intestinal tract [39]. Vitamin E also prevents cancer by inhibiting carcinogens from precursor substances. Whereas, vitamin K play functional role by act as blood clotting factor in liver. The B vitamin folic acid significantly also reduces the risk of neural tube defects (NTDs) like spina bifida in new born babies.

### 3.16. Fiber

Legumes are good source of fiber. Dietary fiber comes from the portion of plants that is not digested by enzymes in the intestinal tract. Bacteria present in lower gut may metabolize this and produce short chain fatty acid. Fiber also reduces body cholesterol level by binding with cholesterol in human gut. High fiber foods can improve serum lipoprotein values, lowers blood pressure and improves blood glucose level for diabetic individuals [35]. Insoluble fiber increases the rate of transient time of wastes material from the gastrointestinal tract.

### 3.17. Oligosaccharides

Indigestible substances especially flatulence induced oligosaccharides ($\alpha$-galactosides) e.g. raffinose, stachyose and verbascose, occur mainly in legume seeds. Oligosaccharides show wide range of physiological properties like anticarcinogenic effect, antidiabetic, anti-cardiovascular and higher rate of mineral absorption that are beneficial to human health [43]. Prebiotic such as Bifidobacteria prevent colon cancer.

### 3.18. Phytic acid

Phytic acid of legumes reduces colon cancer via chelation of iron and suppression of iron-related initiation and promotion of carcinogenesis. Further, it may have potential therapeutic use in cancer due to its property of increasing the activity of natural killer cells associated with suppressed tumor incidence [48, 49].

### 3.19. Vitamins and minerals

Soybean is a better source of vitamins B compared to cereals, although it lacks B12 and vitamin C [50]. Soybean oil also contains tocopherols which are tremendous natural antioxidants. Soybean also contains 5% minerals. It is relatively rich in K, P, Ca, Mg and Fe. Soy ferritin can be extra reasonable quantities of iron [51].

### 3.20. Phytosterols

Phytosterols are natural compounds structurally similar to mammalian cell-derived cholesterol. The best dietary sources of phytosterols are unrefined vegetable oils, seeds, cereals, nuts, and legumes. Phytosterols with potential effects on obesity are diosgenin, campesterol, brassicasterol, sitosterol, stigmasterol, and guggulsterone [52]. High intakes of these compounds can also protect against atherosclerosis and decrease serum TC and LDL-C levels [53]. Their influence on intestinal genes and transcription factors makes phytosterols key regulators in metabolism and cholesterol transport in the expression of liver genes.

### 3.21. Phytoestrogens

Phytoestrogens in food legumes are nonsteroidal phytochemicals quite similar in structure and function to gonadal estrogen hormone. They have antioxidant effects due to their polyphenolic nature, including modulation of steroid metabolism or of enzymes detoxification, interference with calcium transport, and positive effects on lipid and lipoprotein profiles [52]. They offer an alternative therapy for hormone replacement therapy with beneficial effects on the cardiovascular system, and may even improve menopausal symptoms.

### 3.22. Phytochemicals in legumes

Legumes contain, in addition to the health-promoting components (fibers, proteins, resistant starch, and minerals), numerous phytochemicals endowed with useful biological activities [53]. Various studies reveal that selected polyphenols exhibit strong protective actions on many pathological conditions, particularly those triggered by oxidative stress, such as CVD and metabolic disorders [54]. The major sources of dietary polyphenols are cereals, legumes (beans and pulses), oilseeds, fruits, vegetables, and some beverages.

### 3.23. Legume fibers

Legumes are a very good source of dietary fibers. Dietary fibers include resistant starch, non-starch polysaccharides (cellulose, hemicellulose, pectin, gums, and b-glucans), nondigestible oligosaccharides, and lignin. High consumption of soluble fibers is associated with a decrease in serum total cholesterol (TC), in LDL-C, and is inversely correlated with CHD mortality rates [53, 54]. Dietary fibers may also be beneficial against obesity.

### 3.24. Antioxidant activity

The antioxidant properties of food have been studied since reactive oxygen species are widely believed to be involved in many diseases such as cancer, diabetes, autoimmune conditions, various respiratory diseases, eye diseases, and schizophrenia [55, 56]. The antioxidant activity of different dry beans has been assessed by several workers [8–12, 51–57].

### 3.25. Anti-nutritional components (ANC)

Despite the potential nutritional and health-promoting values of legume seeds the presence of anti-nutritional components (ANC) in it, limits its biological value and usage as food. Bean seeds like chick pea, grain beans, lupin, peanut etc. contain a number of antinutritional compounds which can be of proteinous or non-proteinous nature [58]. ANCs found in legume seeds are mainly ubiquitarian, like proteinase inhibitors, lectins, phytates, polyphenols, other are more specific, as some complex glycosides [59]. Legume seeds antinutrients are considered to limit protein and carbohydrate utilization. The negative effects of legume seeds are only observed after the consumption of raw and unprocessed seeds or flour, as normally high temperature during processing inactivates the ANCs [60]. Most of the bean ANCs have an impact on the digestive system, like the inhibition of digestive enzymes (*e.g.* protease inhibitors), impairment of hydrolytic functions and of transport at the enterocyte site (lectins), formation of insoluble complexes (phytates, polyphenols), and the increase of the production of gases in the colon ($\alpha$-galactosides) [58]. The most characterized and commonly found protein inhibitors of legume seeds are trypsin inhibitor of both, Bowman-Birk type and Kunitz type, and $\alpha$-amylase inhibitors like in chickpea [61]. Most of legume species like kidney bean, grain beans, etc. are also a good source of lectins [62]. Legume seeds are also contains a number of non-protein ANCs, like phenolic compounds, saponins, alkaloids, phytates etc. that impair the biological utilization of their nutrients [63].

### 3.26. Legume allergies

The use of legume seeds is being expanded in the food industry due to their excellent nutritional and technological properties. However, legumes have been considered causative agents of allergic reactions through ingestion. Allergy due to consumption of legume seeds for primary sensitization or cross-reactions with other legumes is also a major concern due to the presence of anti-nutritional components. In 2006, Lupin allergy was reported in the literature and there after the European Union included lupin among the allergens that must mandatorily be declared on the foodstuff labels [64]. The main allergens that have been associated with the sensitization to lupin are $\alpha$- and ß-conglutins and, to a lesser extent, $\gamma$- and δ-conglutin [65, 66]. Peanut allergy is one of the most common and severe IgE-mediated reactions to food and is typically lifelong [67]. Lentils and chickpeas have also been reported to cause IgE-mediated hypersensitivity reactions, particularly in pediatric patients [68]. The major lentil allergen was identified as Len c 1 (as a 48-kDa vicilin) [69]. So far, no chickpea allergen has been identified but several IgE-binding bands (10–70 kDa) have been detected by immunoblotting [68]. Total 33 soybean proteins have been identified as allergens (from 7 to 71 kDa) to date [69]. There is need of research to investigate the impact of various treatments in raw legume seeds like roasted peanuts, lentils, chickpeas and soybeans etc. using a serum pool from sensitized patients to find out the possible allergy reaction and its possible overcome.

## 4. Impact of bioactive compounds present in legume seeds on human health

Several bioactive compounds may exhibit a wide range of beneficial effect on human health which may contribute to its nutraceutical property.

### 4.1. Cardiovascular diseases

Legumes help immensely in cholesterol lowering mechanism. There is 22–11% lowering risk of coronary heart disease and cardiovascular disease associated with legume consumption [5]. It has been reported that legume seed consumption lowers LDL cholesterol by partially interrupting the enterohepatic circulation of the bile acids and increasing the cholesterol saturation by increasing the hepatic secretion of cholesterol [15]. Different types of bioactive compounds found in legume seeds such as fibers, oligosaccharides, angiotensin converting enzyme (ACE), vitamins and minerals help in the protection against cardiovascular diseases. It has been examined that 30% neutral detergent fiber of black gram in diet can reduce cholesterol level compared to cellulose on binding with bile acids [31].

### 4.2. Diabetes

Legume seeds play a significant role in the treatment of diabetes. They have high content of fiber and oligosaccharide which help in maintaining the glycemic level in blood. Low glycemic index corresponds to reduce lipid level, insulin and epididymal adipocyte volume in plasma. Hence legume seeds can be used as an anti-diabetic agent [51, 58].

## 4.3. Cancer

Legume seeds contain several nutrients and bioactive compounds like fibers, oligosaccharides, phenolic compounds and antioxidants that show anti-carcinogenic activity. It has been found that adzuki bean (*Vigna angularis*) has differentiation/maturation inducing activity for dendritic cells and apoptosis inducing activity for human leukemia U937 cell. Hence legume seeds help in the treatment of different types of cancer [41]. There is a reduced risk of prostrate cancer, breast cancer, colon cancer and pancreatic cancer on consumption of legume seeds [41, 57].

## 4.4. Hepatotoxicity

Legume seeds are healthy food for liver. γ-aminobutyric acid (GABA) present in legume seeds is a potent hepatoprotective agent [5].

## 4.5. Osteoporosis

Legume seeds being a good source of calcium and protein help to build strong bone and hence reduces the risk of osteoporosis. Isoflavones daidzein and genistein prevents breakdown of bones [51].

## 4.6. Postprandial hyperglycemia

Legume seed husks possess potent anti-oxidant and anti-hyperglycemic activity. It may become an economical natural organic resource for development of functional food/nutraceuticals meant for hyperglycemic individuals. Methanolic extract of seed husk potentially mitigated development of postprandial hyperglycemic spikes and glycemic load close to clinically used drug acarbose [52].

## 4.7. Anti-carcinogenic effects

Various studies have reported results for intakes of pulses and cancer risk; it will be very difficult, using conventional epidemiological tools, to ascertain the quantitative contribution made by pulses to cancer risk however the beneficial effects against cancer of specific and isolated legume components have been carried out since long. Among the potential protective components against cancer which can be present in pulses are included protease inhibitors, saponins, phytosterols, isoflavones and phytates [41, 70].

## 4.8. Weight control and obesity

Dry legumes are claimed to help maintaining a regular body weight, thanks to their great satiety effect, thus limiting the overall food daily intake. Various seed components have been claimed to bring about this effect. According to studies, performed with healthy subjects, showed an increase of the stool weight when they included soybean or pea fibers in the diet [31, 34]. A specific direct action of grain legume alpha-amylase protein inhibitors has been considered for its potential use in the prevention and therapy of obesity and diabetes [34].

### 4.9. Cryptic activity

The bioactivity of peptides which are latent until released from a protein by enzymatic proteolysis, are called cryptic activities [71]. Many kinds of bioactive peptides which might prevent lifestyle-related diseases are released from food proteins after enzymatic digestion. This can occurs during gastrointestinal digestion, germination, fermentation (the proteolytic systems of bacteria can contribute to the liberation of bioactive peptides) or food processing. In this latter case, the wise use of different proteases may bring to the release of different peptides from the same protein source.

It has often been shown that peptides originating from enzymatic hydrolysis of food proteins demonstrate biological effects in various test systems, their activity being due to the ability of inhibiting endogenous enzyme activities or binding to peptide hormone receptors. An immunostimulating peptide isolated from an enzymatic digest of soybean protein prevented alopecia induced by cancer chemotherapy [72].

### 4.10. Hypolipidemic effect

Dyslipidemias are an important risk factor for coronary artery disease. Insulin resistance, a consequence of increased triglyceride and low-density lipoprotein cholesterol (LDL-C) in plasma and decreased high-density lipoprotein cholesterol (HDL-C), is an important risk factor for peripheral vascular disease, stroke and coronary artery disease [73]. It has been shown that long-term feeding with beans decreases cholesterol and low-density lipoprotein (LDL) serum levels in humans, so it seems likely that it can offer protection against cardiovascular diseases [74]. The fiber isolated from *Phaseolus mungo* showed a neutral detergent residue (NDR). It has significant cholesterol lowering activity and increased bile acid excretion in feces [75].

### 4.11. Chronic degenerative diseases

Chronic diseases can be defined as disorders which last for a long time and progress slowly such as heart disease, infarcts, cancer, pulmonary diseases and diabetes. These diseases are the main causes of mortality in the world, accountable for 63% of deaths [76]. Bean consumption has been related to numerous health benefits, such as a decrease in cholesterol levels and cardiac diseases. Beans also offer some protection against cancer, diabetes and obesity, because of their antioxidant, antimutagenic and antiproliferative properties.

### 4.12. Anti-inflammatory agent

BBI proteases from legume crops achieved investigational new drug status by the FDA due to their health-promoting benefits in a denatured form [77]. Since then numerous reports have been published on the potential health-promoting benefits of protease inhibitors, such as their potential use as an anti-inflammatory agent.

### 4.13. Hypertension

ACE causes high blood pressure by converting the biologically inactive angiotensin I to the potent vasoconstrictor angiotensin II, and also inactivates the vasodilator bradykinin [78].

Angiotensin I-converting enzyme (ACE) inhibitor peptides have been isolated from various legumes and has proven to be effective in the prevention and treatment of hypertension.

### 4.14. Glycemic control and diabetes

Clinical studies consistently show that when replacing other carbohydrate-rich foods, bean reduces postprandial glucose elevations in both diabetic and nondiabetic participants [34]. When comparing extreme quintiles, bean (a category that excludes peanuts and soy products) intake was associated with a significant decreased risk of developing diabetes and bean consumption may decrease CVD as well as other diseases by reducing inflammation.

### 4.15. Mortality

Bean consumption has been associated with reduced risk of mortality. Legume intake ranged from 85 g/d in Japan and Greece to a low of only 14 g/d in some segments of the Australian population, legumes were the only foods associated with a reduced risk of mortality [79].

## 5. Conclusions

Legume seeds are nutritious food for the people around the world as it is known to produce richer quality of proteins (around 20–40% protein) than cereals. Apart from protein, it is also composed of carbohydrates, fiber, amino acids, micronutrients including several vitamins and minerals. Legumes are also well known for the presence of different bioactive compounds such as saponins, tannins, flavonoids, isoflavones, lectins, phytic acid etc. which is important for its nutraceutical property and provides beneficial effects on human health as well as helps in the prevention or treatment of certain diseases such as cardiovascular diseases, diabetes, digestive tract diseases, overweight, obesity, cancer, etc. There are many more underutilized food legume seeds that may be a potential source of nutraceutical food. The importance of legumes in human diet is expected to increase in the near future in order to meet the demand for protein and other nutrients in the increasing world population and also to reduce the risk related to animal food source consumption.

## Acknowledgements

Authors express thanks to North Eastern Hill University, Tura Campus, Meghalaya, India.

## Conflict of interest

There is no conflict of interest among the authors.

## Author details

Arindam Barman*, Chinky M. Marak, Rituparna Mitra Barman and Cheana S. Sangma

*Address all correspondence to: arindamnehu@gmail.com

North Eastern Hill University, Meghalaya, India

## References

[1]  Costa JP. A current look at nutraceuticals-key concepts and future prospects. Trends in Food Science and Technology. 2017;**62**:68-78. DOI: 10.1016/j.tifs.2017.02.010

[2]  Morris B. Bio-functional legumes with nutraceutical, pharmaceutical and industrial uses. Economic Botany. 2003;**57**:254-261. DOI: 10.1017/S1479262113000397

[3]  Vadivel V, Patel A, Biesalski HK. Effect of traditional processing methods on the anti-oxidant, $\alpha$-amylase and $\alpha$-glucosidase enzyme inhibition properties of Sesbania sesban Merrill seeds. CyTA: Journal of Food. 2012;**10**:128-136. DOI: 10.1080/19476337.2011.601427

[4]  Scarafoni A, Magni C, Duranti M. Molecular nutraceutics as a mean to investigate the positive effects of legume seed proteins on human health. Trends in Food Science & Technology. 2007;**18**:454-463. DOI: 10.1016/j.tifs.2007.04.002

[5]  Shweta KM, Rana A. Bioactive components of Vigna species: Current prospective. Bulletin of Environment, Pharmacology and Life Sciences. 2017;**6**:1-13

[6]  González-Montoya M, Cano-Sampedro E, Mora-Escobedo R. Bioactive peptides from legumes as anticancer therapeutic agents. International Journal of Cancer and Clinical Research. 2017;**4**:1-10. DOI: 10.23937/2378-3419/1410081

[7]  Xu BJ, Chang SKC. A comparative study on phenolic profiles and antioxidant activities of legumes as affected by extraction solvents. Journal of Food Science. 2007;**72**:159-166. DOI: 10.1111/j.1750-3841.2006.00260.x

[8]  Siddhuraju P, Becker K. The antioxidant and free radical scavenging activities of pro-cessed cowpea (*Vigna unguiculata* (L.) Walp.) seed extracts. Food Chemistry. 2007;**101**: 10-19. DOI: 10.1016/j.foodchem.2006.01.004

[9]  Pei-Yin L, Hsi-Mei L. Bioactive compounds in legumes and their germinated products. Journal of Agricultural and Food Chemistry. 2006;**54**:3807-3814. DOI: 10.1021/jf060002o

[10]  Bhattacharya S, Malleshi NG. Physical, chemical and nutritional characteristics of pre-mature-processed and matured green legumes. Journal of Food Science and Technology. 2012;**49**:459-466. DOI: 10.1007/s13197-011-0299-y

[11]  Loganayaki N, Siddhuraju P, Manian S. A comparative study on in vitro antioxidant activity of the legumes *Acacia auriculiformis* and *Acacia ferruginea* with a conventional legume. CyTA: Journal of Food. 2011;**9**:8-16. DOI: 10.1080/19476330903484216

[12]  Clemente A, Olias R. Beneficial effects of legumes in gut health. Current Opinion in Food Science. 2017;**14**:32-36. DOI: 10.1016/j.cofs

[13]  Sefatie RS, Fatoumata T, Eric K, Shi YH, Guo-wei L. In vitro antioxidant activities of protein hydrolysate from germinated black soybean (*Glycine max* L.). Advance Journal of Food Science and Technology. 2013;**5**:453-459

[14]  Tiwari AK, Abhinay B, Babu KS, Kumar DA, Zehra A, Madhusudana K. Pigeon pea seed husks as potent natural resource of anti-oxidant and anti-hyperglycaemic activity. International Journal of Green Pharmacy. 2013;**9**(4-5):36-49.36-49. DOI: 10.4103/0973-8258.120247

[15]  Shahwar D, Mohsin T, Bhat MYK, Chaudhary S, Aslam R. Health functional compounds of lentil (*Lens culinaris* Medik): A review. International Journal of Food Properties. 2017;**20**:1-15. DOI: 10.1080/10942912.2017.1287192

[16]  Chibbar RN, Ambigaipalan P, Hoover R. Molecular diversity in pulse seed starch and complex carbohydrates and its role in human nutrition and health. Cereal Chemistry. 2010;**87**:342-352. DOI: 10.1094/CCHEM-87-4-0342

[17]  Geervani P. Utilization of chickpea in India and scope for novel and alternative uses. In: Proceedings of a Consultants Meeting; 27-30 March 1989; Patancheru, India: ICRISAT; 1991. pp. 47-54

[18]  Akihisa T, Yasukawa K, Yamaura M. Triterpene alcohol and sterol ferulates from rice bran and their anti-inflammatory effects. Journal of Agricultural and Food Chemistry. 2000;**48**:2313-2319. DOI: 10.1021/jf000135o

[19]  Murty CM, Pittaway JK, Ball MJ. Chickpea supplementation in an Australian diet affects food choice, satiety and bowel function. Appetite. 2010;**54**:282-288. DOI: 10.1016/j.appet.2009.11.012

[20]  Duranti M, Consonni A, Magni C, Sessa F, Scarafoni A. The major proteins of lupin seed: Characterisation and molecular properties for use as functional and nutraceutical ingredients. Trends in Food Science & Technology. 2008;**19**(12):624-633. DOI: 10.1016/j.tifs.2008.07.002

[21]  Arnoldi A, Boschin G, Zanoni C, Lammi C. The health benefits of sweet lupin seed flours and isolated proteins. Journal of Functional Foods. 2015;**18**:550-563. DOI: 10.1016/j.jff.2015.08.012

[22]  Whiteman H: Eating Nuts, Peanuts Daily could Lower Death Risk from Cancer, Other Diseases [Internet]. 2015. Available from: https://www.medicalnewstoday.com/articles/295124.php [Accessed: 2018-05-11]

[23]  Crepon K, Marget P, Peyronnet C, Carrouee B, Arese P, Duc G. Nutritional value of faba bean (*Vicia faba* L.) seeds for feed and food. Field Crops Research. 2010;**115**:329-339. DOI: 10.1016/j.fcr.2009.09.016

[24]  Smartt J. Gene pools in grain legumes. Economic Botany. 1984;**38**(1):24-35. DOI: 10.1007/bf02904413

[25] Bressani R, Gómez-Brenes RA, Elías LG, Hobart. Nutritional quality of pigeon pea protein, immature and ripe, and its supplementary value for cereals. Archivos Latinoamericanos de Nutrición. 1986;**36**(1):108-116

[26] Pande S, Platel K, Srinivasan K. Antihypercholesterolaemic influence of dietary tender cluster beans (*Cyamopsis tetragonoloba*) in cholesterol fed rats. The Indian Journal of Medical Research. 2012;**135**(3):401-406

[27] Vedavyas R, Niveditha, Kandikere R, Sridhar. Antioxidant activity of raw, cooked and Rhizopus oligosporus fermented beans of Canavalia of coastal sand dunes of Southwest India. Journal of Food Science and Technology. 2012;**51**(11):3253-3260. DOI: 10.1007/s13197-012-0830-9

[28] Shashank A, Tidke D, Ramakrishna S, Kiran G, Kosturkova GA, Ravishankar. Nutraceutical potential of soybean: Review. Asian Journal of Clinical Nutrition. 2015;**7**:22-32. DOI: 10.3923/ajcn.2015.22.32

[29] Peter B, Kaufman JA, Duke BH, Boik J, Hoyt JE. A comparative survey of leguminous plants as sources of the isoflavones, genistein and daidzein: Implications for human nutrition and health. The Journal of Alternative and Complementary Medicine. 1997;**3**:7-12. DOI: 10.1089/acm.1997.3.7

[30] Prakash D, Upadhyay G, Singh BN, Singh HB. Antioxidant and free radical-scavenging activities of seeds and agri-wastes of some varieties of soybean (*Glycine max*). Food Chemistry. 2006;**104**:783-790. DOI: 10.1016/j.foodchem.2006.12.029

[31] Guillon F, Champ MJ. Carbohydrate fractions of legumes: Uses in human nutrition and potential for health. The British Journal of Nutrition. 2002;**88**:5293-5306. DOI: 10.1079/BJN2002720

[32] Colonna P, Mercier C. Gelatinization and melting of maize and pea starches with normal and high amylose genotypes. Phytochemistry. 1985;**24**:1667-1674. DOI: 10.1016/S0031-9422(00)82532-7

[33] Englyst HN, Kingman SM, Cummings JH. Classification and measurement of nutritionally important starch fractions. European Journal of Clinical Nutrition. 1992;**46**:S33-S50. DOI: 10.12691/jfnr-1-6-7

[34] Kirsten R, Heintz B, Nelson K, Hesse K, Schneider E, Oremek G, Nemeth N. Polyenylphosphatidylcholine improves the lipoprotein profile in diabetic patients. International Journal of Clinical Pharmacology and Therapeutics. 1993;**32**:53-56. DOI: 10.1194/jlr.M400438

[35] Anderson JW, Major AW. Pulses and lipaemia, short- and long-term effect: Potential in the prevention of cardiovascular disease. The British Journal of Nutrition. 2002;**88**:5263-3271. DOI: 10.1079/BJN2002716

[36] Shaw BP, Bera CH. A preliminary clinical study to cultivate the effect of vigorex-SF in sexual disability patients. Indian Journal of Internal Medicine. 1993;**3**:165-169

[37] Rajendran V, Joseph T, David J. Reappraisal of dopaminergic aspects *Mucuna pruriens* and comparative profile with L-dopa on cardiovascular and central nervous system in animals. Indian Drugs. 1996;**33**:65-72. DOI: 10.1158/1541-7786.MCR-13-0531

[38] Hussain G, Manyam BV. Mucuna pruriens proves more effective than L-DOPA in Parkinson's disease animal model. Phytotherapy Research. 1997;**11**:419-423. DOI: 101007/s11240-010-9804-7

[39] Tharakan B, Dhanasekaran M, Mize-Berge J, Manyam BV. Anti-Parkinson botanical *Mucuna pruriens* prevents levodopa induced plasmid and genomic DNA damage. Phototherapy Research. 2007;**21**:1124-1126. DOI: 10.13140/RG.2.1.1586.1367

[40] Liaudet L, Feihl F, Rosselet A, Markert M, Hurni JM, Perret C. Beneficial effects of L-canavanine, a selective inhibitor of inducible nitric oxide synthase, during rodent endotoxaemia. Clinical Science. 1996;**90**:369-377. DOI: 10.1042/cs0900369

[41] Swaffar DS, Ang CY, Desai PB, Rosenthal GA. Inhibition of the growth of human pancreatic cancer cells by the arginine antimetabolite L-canavanine. Cancer Research. 1994;**54**: 6045-6048

[42] Morris JB. Legume genetic resources with novel value added industrial and pharmaceutical use. In: Janick J, editor. Perspectives on New Crops and New Uses. Alexandria, Virginia: ASHS Press; 1999. pp. 196-201

[43] Rodrigues BF, Torne SG. A chemical study of seeds in three *Canavalia* species. Tropical Science. 1991;**31**:101-103

[44] Ruediger H, Gabius HJ. Plant lectins: Occurrence, biochemistry, functions and applications. Glycoconjugate Journal. 2001;**18**:589-613

[45] Lee MC, Damjanov I. Anatomic distribution of lectin-binding sites in mouse testis and epididymis. Differentiation. 1984;**27**:74-81. DOI: 10.1111/j.1432-0436.1984.tb01410.x

[46] Jerusalmi A, Farlow SJ, Sano T. Use of lectin as an anchoring agent for adenovirus-microbead conjugates: Application to the transduction of the inflamed colon in mice. Gene Therapy and Molecular Biology. 2006;**10**:223-232

[47] Cuatrecasas P, Tell GPE. Insulin-like activity of concanavalin A and wheat germ agglutinin-direct interactions with insulin receptors. Proceedings of the National Academy of Science. 1973;**70**:485-489. DOI: 10.1073/pnas.70.2.485

[48] Scarafoni A, Kumar J, Magni C, Sironi E, Duranti M. Biologically active molecules and nutraceutical properties of legume seeds. In: Proceeding of the Fourth International Food Legumes Research Conference (IFLRC-IV); 18-22 October 2005; New Delhi. India: ISGPB; 2007. pp. 18-22

[49] Roy F, Boye JI, Simpson BK. Bioactive proteins and peptides in pulse crops: Pea, chickpea and lentil. Food Research International. 2010;**43**:432-442. DOI: 10.1016/j.foodres.2009.09.002

[50] Liu KS. Chemistry and nutritional value of soybean components. In: Liu KS, editor. Soybeans: Chemistry, Technology and Utilization. New York, USA: Chapman and Hall; 1997. pp. 25-113

[51] Sugano M. Soy in Health and Disease Prevention. Boca Raton: CRC Press; 2006. 328 p

[52] Racette SB, Spearie CA, Phillips KM, Lin X, MA L, MS, Ostlund RE. Phytosterol-deficient and high-phytosterol diets developed for controlled feeding studies. Journal of the American Dietetic Association. 2009;**109**(12):2043-2051. DOI: 10.1016/j.jada.2009.09.009

[53] Tiwari AT, Abhinay B, Babu KS, Kumar DA, Zehra A, Madhusudana K. Pigeon pea seed husks as potent natural resource of antioxidant and anti hyperglycaemic activity. International Journal of Green Pharmacy. 2013;**7**:252-257. DOI: 10.4103/0973-8258.120247

[54] Bouchenak M, Lamri-Senhadji M. Nutritional quality of legumes, and their role in cardio-metabolic risk prevention. Journal of Medicinal Food. 2013;**16**(3):185-198. DOI: 10.1089/jmf.2011.0238

[55] Campos-Vegaa R, Loarca-Piñaa G, Oomah BD. Minor components of pulses and their potential impact on human health. Food Research International. 2010;**43**:461-482. DOI: 10.1016/j.foodres.2009.09.004

[56] Cai Y, Luo Q, Sun M, Corke H. Antioxidant activity and phenolic compounds of 112 traditional Chinese medicinal plants associated with anticancer. Life Sciences. 2004;**74**: 2157-2184. DOI: 10.1016/j.lfs.2003.09.047

[57] Kandikere R, Sridhar VR, Niveditha. Wild Legume Canavalia Cathartica—An Overview on Nutritional and Bioactive Potential. India: Nova Science Publishers; 2001. 574 p

[58] Krupa U. Main nutritional and antinutritional compounds of bean seeds—A review. Polish Journal of Food and Nutrition Sciences. 2008;**58**(2):149-155. DOI: 10.1016/0963-9969(93)90069-U

[59] Fernandes AO, Banerji AP. Long-term feeding of field bean protein containing protease inhibitors suppresses virus-induced mammary tumors in mice. Cancer Letters. 1997;**116**: 1-7. DOI: 10.1016/S0304-3835(98)00090-1

[60] Clemente A, MacKenzie DA, Jonson IT, Domoney C. Investigation of legume seed protease inhibitors as potential anticarcinogenic proteins. In: Proceedings of the Fourth International Workshop on Antinutritional Factors in Legume Seeds and Oilseeds; 8-10 March 2004; EAA Publications. Wageningen; 2004. pp. 137-141

[61] Leterme P. Recommendations by health organizations for pulse consumption. British Journal of Nutrition. 2002;**88**(3):239-242. DOI: 10.1079/BJN2002712

[62] Deshpande SS, Nielsen SS. *In vitro* enzymatic hydrolysis of phaseolin, the major storage protein of *Phaseolus vulgaris* L. Journal of Food Science. 1987;**52**:1326-1329. DOI: 10.1111/j.1365-2621.1987.tb14074.x

[63] European Commission. Directive 2006/142/EC of 22 December 2006, amending annex IIIa of directive 2000/13/EC of the European Parliament and of the council listing the ingredients which must under all circumstances appear on the labelling of foodstuffs. Official Journal of the European Union. 2006;**368**:110-111

[64] Ballabio C, Pen E, Uberti F, Fiocchi A, Duranti M, Magni C, Restani P. Characterization of the sensitization profile to lupin in peanut-allergic children and assessment of cross-reactivity risk. Pediatric Allergy and Immunology. 2013;24:270-275. DOI: 10.1111/pai.12054

[65] Jimenez-Lopez JC, Foley RC, Breard E, Clarke VC, Lima-Cabelloa E, Floridoe JF, Singh KB, Alchea JD, PMC S. Characterization of narrow-leaf lupin (*Lupinus angustifolius* L.) recombinant major allergen IgE-binding proteins and the natural β-conglutin counterparts in sweet lupin seed species. Food Chemistry. 2018;244:60-70. DOI: 10.1016/j. foodchem.2017.10.015

[66] Martínez M, Ibanez MD, Fernandez-Caldas E, Maranon F, Rosales MJ, Laso MT. Specific IgE levels to *Cicer arietinum* (chickpea) in tolerant and nontolerant children: Evaluation of boiled and raw extracts. International Archives of Allergy and Immunology. 2000;121: 137-143. DOI: 10.1159/000024309

[67] Patil SP, Niphadkar PV, Bapat MM. Chickpea: A major food allergen in the Indian subcontinent and its clinical and immunochemical correlation. Annals of Allergy, Asthma & Immunology. 2001;87:140-145. DOI: 10.1016/S1081-1206(10)62209-0

[68] Sanchez-Monge R, Pascual CY, Diaz-Perales A, Fernández-Crespo J, Martín-Esteban M, Salcedo G. Isolation and characterization of relevant allergens from boiled lentils. The Journal of Allergy and Clinical Immunology. 2000;106:955-961. DOI: 10.1067/mai. 2000.109912

[69] Wilson S, Blaschek K, Gonzalez de Mejia E. Allergenic proteins in soybean: Processing and reduction of P34 allergenicity. Nutrition Reviews. 2005;63:47-58

[70] Mathers JC. Pulses and carcinogenesis: Potential for the prevention of colon, breast and other cancers. The British Journal of Nutrition. 2002;8:3273-3279. DOI: 10.1079/BJN 2002717

[71] Meisel H, Bockelmann W. Bioactive peptides encrypted in milk proteins: Proteolytic activation and thropho-functional properties. Antonie Van Leeuwenhoek. 1999;76:207-215. DOI: 10.1023/A:1002063805780

[72] Yoshikawa M, Fujita H, Matoba N, Takenaka Y, Yamamoto T, Yamauchi R, Tsuruki H, Takahata K. Bioactive peptides derived from food proteins preventing lifestyle-related diseases. BioFactors. 2002;12:143-146. DOI: 10.1080/10408398.2012.753866

[73] Jellinger PS, Smith DA, Mehta AE, Ganda O, Handelsman Y, Rodbard HW, Seibel JA, AACE task force for management of dyslipidemia and prevention of atherosclerosis. American Association of Clinical Endocrinologists' Guidelines for management of dyslipidemia and prevention of atherosclerosis. Endocrine Practice. 2012;18(Suppl 1):1-78. DOI: 10.4158/EP.18.S1.1

[74] Marzolo MP, Amigo L, Nervi F. Hepatic production of very low density lipoprotein, catabolism of low density lipoprotein, biliary lipid secretion, and bile salt synthesis in rats fed a bean (*Phaseolus vulgaris*) diet. Journal of Lipid Research. 1993;34:807-814

[75] Thomas M, Leelamma S, Kurup PA. Effect of blackgram fiber (*Phaseolus mungo*) on hepatic hydroxymethylglutaryl-CoA reductase activity, cholesterogenesis and cholesterol degradation in rats. Journal of Nutrition. 1983;**113**:1104-1108. DOI: 10.4172/2161-0444.1000241

[76] World Health Organization. Health Issues: Chronic Diseases [Internet]. 2013. Available from: http://www.who.int/topics/chronic_diseases/es/ [Accessed: 2018-04-13]

[77] Kennedy AR, Szuhaj BF, Newberne PM, Billings PC. Preparation and production of a cancer chemopreventive agent, Bowman-Birk inhibitor concentrate. Nutrition and Cancer. 1993;**19**:281-302. DOI: 10.1080/01635589309514259

[78] Wong JH, Ng TB. Sesquin, a potent defensin-like antimicrobial peptide from ground beans with inhibitory activities toward tumor cells and HIV-1 reverse transcriptase. Peptide. 2005;**26**:1120-1126. DOI: 10.1016/j.peptides.2005.01.003

[79] Lousuebsakul-Matthews V, Thorpe DL, Knutsen R, Beeson WL, Fraser GE, Knutsen FS. Legumes and meat analogues consumption are associated with hip fracture risk independently of meat intake among Caucasian men and women: The Adventist Health Study-2. Public Health Nutrition. 2014;**17**(10):2333-2343. DOI: 10.1017/S1368980013002693

# Two Sides of the Same Coin: The Impact of Grain Legumes on Human Health: Common Bean (*Phaseolus vulgaris* L.) as a Case Study

Elsa Mecha, Maria Eduardo Figueira,
Maria Carlota Vaz Patto and
Maria do Rosário Bronze

Additional information is available at the end of the chapter

http://dx.doi.org/10.5772/intechopen.78737

## Abstract

Data from Food and Agriculture Organization indicate the worrying scenario of severe food insecurity in the world and the contrasting high prevalence of obesity (13% of the world adult population) in both developing and developed countries. Sustainable agriculture systems with increased inclusion of grain legume species and the boosting of public awareness about legume importance on diet should be a priority issue to eradicate malnutrition and promote public health. However, grain legume production and consumption are in constant state of decline, especially in the European Union. Assigned as the "poor man's meat", "promoters of flatulence", or incorrectly classified as "starchy foods", grain legumes have a negative image in modern societies. In fact, legumes represent an important source of protein, fiber, vitamins (e.g. folate) and minerals (e.g. magnesium). Moreover, legumes are rich in bioactive compounds (e.g. phenolic compounds, protease and $\alpha$-amylase inhibitors) acting as a "double-edged sword" in human health. They may impair nutrients availability exerting at the same time beneficial biological activities in lipid profile, inflammation, glycaemia and weight. The present chapter is focused on the advantages of a legume-rich diet for health promotion at a global scale, reviewing legume nutritional and bioactive compounds, with particular emphasis on common bean.

**Keywords:** grain legumes, nutritional value, bioactive compounds, health benefits

# 1. Introduction

Grain legumes have been neglected, regardless of their potential to ensure nutrition and food security. Nutritionally rich in protein, fiber, carbohydrates, vitamins and minerals, grain legumes are key dietary components to eradicate hunger, as well as, malnutrition [1].

The ignorance regarding grain legume nutritional composition and food preparation techniques, allied with the negative image of legumes in modern societies, contributes to decrease legumes' consumption. Besides nutrients, legumes are also a rich source of bioactive compounds which can act as a "double-edged sword", since they can impair nutrients' bioavailability (as anti-nutritional factors), acting simultaneously, as health promoting compounds in the prevention of non-communicable diseases (e.g. cardiovascular diseases, inflammatory diseases and cancer) [2]. In order to balance negative and positive effects of these bioactive compounds, crops diversity should be preserved and characterized to give valid information to breeders and molecular biologists, who can manipulate the levels of these compounds through the selection of interesting varieties.

The present chapter aims to give a general overview of the current state of the art of grain legume production, consumption and impact on world food security. It also shows the nutritional value and the bioactive composition considering some *in vitro*, *in vivo* and epidemiological studies conducted to analyze the potential health benefits associated with legumes consumption.

# 2. Legumes diversity

Legumes are dicotyledons plants, which belong to Leguminosae or Fabaceae family, with edible seeds developed in pods. By definition, it includes the fresh legumes, pulses and the seeds with high fat content (e.g. soybeans and peanuts). Pulses, also known as grain legumes, refer only to the dried seeds with virtually no fat, which excludes the fresh legumes, soybeans and peanuts. Common bean (*Phaseolus vulgaris* L.), pea (*Pisum sativum* L.), faba beans (*Vicia faba* L.), chickpea (*Cicer arietinum* L.), lentils (*Lens culinaris* L.) and grass pea (*Lathyrus sativus* L.) are examples of legumes well-adapted to several regions of the world, from semi-arid, subtropical to temperate areas.

The wild form of *P. vulgaris* is originally from Mesoamerica (which extends from northern Mexico to Colombia). Since its expansion, two independent domestication centers were formed in Mesoamerica and Andes (from southern Peru to northwestern Argentina) [3].

In Europe, particularly in Portugal [4], Spain, Italy and central-northern Europe, common bean germplasm derives mostly from the Andean domestication center (67%) and in the Eastern Europe, there is a higher predominance of the Mesoamerican type [3].

Despite the large genetic diversity in grain legume seeds held in gene banks, the genetic resources are not intensively used in breeding programs. Preservation, characterization and evaluation of the genetic variability, in what concerns agronomic performance and quality traits, is a useful approach to ensure *in situ* conservation and future breeding programs to cope with consumers' demands and environmental challenges [5].

# 3. Legumes production and consumption

Diversifying agriculture, instead of adopting an intensive specialized production system, is one of the goals to achieve a sustainable development. Grain legumes bring diversity, nutrient supply and disease control to cropping systems. In opposition to the American continent, Africa, Asia and Oceania, in the European Union, common bean production decreased drastically (−80.42%) between 1961 (817,000 tonnes) and 2013 (160,000 tonnes) [6]. During this period of time, there was a shift in land use toward an intensive cereals production [6], which contributed to the Europeans' dependence in imported grain legumes, compromising sustainability of the actual food farming system. Parallel to the decrease in common bean production data from FAOSTAT, relative to food balance, also indicate, in European Union (EU), a dramatic decrease on its consumption from 1.5 kg/capita/ year (1961) to 0.78 kg/capita/year (2013) [6].

Several factors related to crop productivity, government policies and consumers' preferences can explain the reduced investment of European farmers in grain legumes production. The promotion of breeding programs to increase genetic diversity and the development of more attractive varieties adapted to the local growing conditions and to the consumers' demands (high quality varieties) must be pursued.

## 3.1. Food security

The Food and Agriculture Organization (FAO) of the United Nations declared 2016 as the International year of Pulses focusing on hunger and malnutrition eradication [7]. According to the second sustainable development goal of FAO, by 2030, countries should "end hunger", adopt sustainable agriculture systems and provide food security to all population [8]. Several factors can affect food security worldwide: extreme weather events (e.g. droughts, floods and hurricanes), conflicts with violence affecting rural areas and economic recessions with increased unemployment [8]. Worrying data from FAO indicate that, in 2016, 815 million people suffer from chronic food deprivation and around 698 million people from severe food insecurity [8].

To avoid the financial pressure of malnutrition on health care systems and the economic burden of the co-morbidities related with malnutrition, governments should support sustainable agriculture practices with inclusion of legumes in cropping systems and subsidies to small farmers, especially in low- and middle-income countries dependent on agriculture [9]. Nutritional initiatives to eradicate malnutrition and protein deficiency should include public awareness about inclusion of vegetable protein in daily diet [8].

# 4. Nutritional value

Legumes are within the food items with a high nutrient value (330 ± 217 kcal/100 g) for a low cost value (0.26 ± 0.22 $/serving) [10].

Grain legumes are distinguished as a rich source of vegetable protein, soluble and insoluble fiber, resistant starch, micronutrients (minerals and vitamins) and several bioactive compounds [11]. When complemented with the cereals' protein, grain legumes can be consumed as a sustainable alternative to animal protein. Despite of the American Cancer Society, the Centers for Disease Control and Prevention and the US Dietary guidelines who classify beans as vegetables, many consumers continue to associate grain legumes to starchy foods, like rice, pasta and tubers [12]. The major differences between legumes and starchy food (cereals) are related with macro- and micronutrients composition.

## 4.1. Macronutrients

The macronutrients should be provided by diet in large amounts to supply the energy and the molecular units that sustain the basal metabolism, physical activity, growth, pregnancy and lactation. The carbohydrates' contribution to total food energy is higher in cereals than in beans and there is an inverse situation for the protein contribution, with beans showing higher protein content than cereals [13].

### 4.1.1. Protein

In legumes, proteins are stored in the parenchyma cells of cotyledons and are classified according to their solubility in different solvents as albumins, water extractable, globulins, extractable in salt solutions, prolamins, extractable in aqueous alcohol and glutelins extractable in weak acid/alkaline solutions. In common bean, globulins are the most predominant fraction of storage proteins (54–79%), followed by albumins (12–30%), glutelins (20–30%) and prolamins (2–4%). The most abundant globulin in common bean is the phaseolin (40–50% of the total globulins) [14].

The structural units of the proteins known as amino acids can be classified as essential and non-essential. The essential ones must be necessarily provided by diet. If some of the eight essential amino acids is lacking, the missing one is named as a "limiting amino acid". In legumes, the limiting amino acids are sulfur-containing amino acids (methionine and cysteine) and in cereals lysine is the limiting one. In order to increase the protein quality of legumes and cereals, both food items must be combined in a daily diet to provide all the essential amino acids and to prevent protein malnutrition [15].

The presence of anti-nutritional factors (trypsin inhibitors, phytic acid and tannins) in grain legumes, detailed below in this chapter, and the processing method used before consumption can influence protein digestibility and protein quality [16].

### 4.1.1.1. Legume proteins as potential allergens

As a rich protein source, legumes may cause allergenic reactions. More than 90% of the food allergies are caused by proteins of vegetable and animal origin [17]. Genetic factors and exposure to new allergenic food products, early in life, can explain the immune response of some individuals to one or more food proteins [18].

In developed countries, more than 6% of the children and around 4% of the adults have food allergies [19]. In developing countries and emerging economies (e.g. Brazil, China and India), the prevalence of food allergies is misreported and under-diagnosed [20].

The food allergy induced by legumes is an IgE immune reaction, characterized by activation of Th2-type lymphocytes [21]. In sensitized individuals, mild (cutaneous rash, diarrhea, vomiting, abdominal pain, hypotension, arrhythmia, repetitive cough, tongue swelling, angioedema, rhinitis and asthma) to severe threatening-life symptoms can occur. The most severe reactions, rarely reported with pulses, include anaphylaxis and death [17].

Since legumes share common antigen determinants (epitopes) with other plants, the risk of an allergenic reaction, in sensitized individuals, increases if cross-reactive foods were not eliminated from diet/environment. For example, pea and common beans have cross-reactivity with pollens of *Olea europaea*, *Lolium perenne* and *Betula alba* [22]. In kidney bean, the major allergens were identified as defense proteins against biotic stress (lectin and α-amylase inhibitor), storage proteins (phaseolin) and stress tolerant proteins (late embryogenesis abundant (LEA) protein). These proteins also showed cross-reactivity with other legumes such as peanut and pigeon pea [18].

To prevent the development of food allergies, the pediatric nutrition authorities recommend exclusive breastfeeding until 6 months of age. Legumes and protein-rich foods (e.g. meat, egg, milk and yoghurt) should be only introduced at the age of 6–8 months [23]. At the agriculture level, promising strategies involving the breeding of crop varieties with reduced content of allergenic proteins are being put into action. Nevertheless, the development of such crops represents a challenge for farmers, who need to deal with compromised plant feasibility [24] and does not represent the appropriate strategy for consumers with severe allergies, since immune reactivity to legumes may occur, even with minimum quantity of allergens. In these patients, the clinical approach to manage allergies should focus on the patients' awareness of a list of food items that must be avoided, and on a personalized nutritional intervention with indication of nutritive food alternatives.

### 4.1.2. Carbohydrates

Legume carbohydrates include starch, fiber and oligosaccharides.

### 4.1.2.1. Starch

Starch represents the main carbohydrate reserve (22–45% of total carbohydrates) in legume seeds and is used by the plant as a source of glucose and energy [25]. Chemically, it is composed by two types of polymers: the amylose and the amylopectin. Amylopectin is a highly branched polymer characterized by a linear chain of glucose moieties linked by α-1,4-glycosidic bonds with several smaller glucose chains at α-1,6 positions. Amylose is a long unbranched linear chain of α-1,4-glucans. A comparative study of the starch structure of a legume (e.g. chickpea) and a cereal (e.g. wheat) revealed the higher content of amylose in chickpea's starch [26]. Starches with high amylose content have low glycemic index and therefore can be more adequate to type 2 *diabetes mellitus* populations [27].

### 4.1.2.2. Dietary fiber

Dietary fiber include the total non-starch polysaccharide (NSP), divided into soluble and insoluble NSP, resistant starch and fructooligosaccharides. Soluble fiber is defined as the fermentable fiber with prebiotic action. The insoluble fiber is poorly fermented and has a bulking function in colon [28]. Compared with cooked corn, cooked beans have higher content of dietary fiber (2.4/100 g in corn against 6.3–10.4/100 g in cooked beans) [13].

Besides total dietary fiber, legumes are also a rich source of resistant starch, which is defined as a portion of starch that passes through the duodenum and jejunum without being digested [28]. In colon, resistant starch is fermented, by the local microbiota, into several products, including short-chain fatty acids (acetate, propionate and butyrate), which are responsible to maintain gut integrity, improve intestinal microflora, reinforce immune system preventing intestinal colonization by pathogens, improve blood lipid profile by reducing plasma triglycerides and LDL cholesterol, control satiety by increasing the secretion of satiety hormones and contribute to prevent several diseases from allergies and autoimmune diseases to bowel cancer [29, 30]. Legumes show higher levels of resistant starch (e.g. 4.3% in kidney beans) than cereals (e.g. 1.4% in rice) and tubers (e.g. 1.8% in potato) in a dry weight basis [31].

### 4.1.2.3. Fructooligosaccharides

Grain legumes are particularly rich in oligosaccharides such as raffinose, stachyose and verbascose, which are likely to be fermented by colonic bacteria. As a consequence of bacterial fermentation, rectal gas is produced, which may be responsible for abdominal discomfort, bloating and flatulence. Since individual gas production is dependent on the individual microflora composition and consumption habits, beans are not necessarily responsible for increased flatulence [32].

Similarly to resistant starch, the colonic fermentation of oligosaccharides is also responsible for the production of short-chain fatty acids, acetate, propionate and butyrate, related to several health benefits [33]. To control the flatulence and reduce the content of oligosaccharides in legumes, many populations, especially in Asia and Africa, consume fermented legumes as an interesting nutritive food alternative [34].

### 4.1.3. Lipids

Lipids represent 2–21% of the macronutrients present in legumes [35]. The content in the different fatty acids is quite variable among the different legume species. By increasing order of the monounsaturated fatty acid (oleic acid) content, common bean has the low amount (5.1–17.2%) followed by lentils (23.5–39.6%), faba beans (25.2–32.4%), peas (26.3–36%) and chickpeas (31.4–34.8%) [36]. However, common beans are particularly rich in polyunsaturated fatty acids (PUFAS), 48.4–68.7% of the lipid content, revealing an higher content of linolenic acid (9,12,15-(Z,Z,Z)-octadecatrienoic acid or C18:3, n-3) than linoleic acid (9,12-(Z,Z)-octadecadienoic acid or C18:2, n-6), ratio n6/n3 between 0.5 and 0.9, which is an indication of the common beans' protective effect against degenerative diseases, such as cardiovascular diseases and inflammatory diseases [36, 37].

### 4.2. Micronutrients

Contrarily to macronutrients, micronutrients are required, by human body, in small amounts performing crucial physiological roles (e.g. metabolism, hormone and enzyme synthesis, immune homeostasis and cell division). Legumes are particularly rich in B-complex vitamins, folate, vitamin E and minerals such as iron, calcium, phosphorus, magnesium, potassium, zinc, copper and selenium [38]. In low- and middle-income countries, highly dependent on legume proteins, the malnutrition by iron deficiency is one of the major worrying public health issues [8]. Although the iron content of a vegetarian diet may be equal to the iron content of a mixed diet, in a non-vegetarian diet, with red meat, the heme iron, mostly present in the form of hemoglobin and myoglobin (10–12% of the total iron) [39] can be absorbed at a rate of 5–35% in the gut. However, in a vegetarian diet (rich in legumes, vegetables and cereals) where the main form of iron is the nonheme, the intestinal absorption decreases to 2–10% [40].

In countries where legumes are staple food products, consumption of biofortified legumes with iron and other micronutrients, such as zinc, with sources of vitamin C can be a solution for micronutrient malnutrition. The fortification of bean varieties with iron is currently a common practice in several countries, such as Rwanda, Uganda, Democratic Republic of Congo and Brazil, in order to control women and childhood iron deficiencies [41].

# 5. Bioactive compounds

Additional to the nutritional value of legumes in human health, legumes are also a rich source of several minor bioactive compounds (e.g. lectins, enzymatic inhibitors, saponins, phytates, oligosaccharides, sterols and phenolic compounds), whose presence has been linked to several nutraceutical properties [42].

## 5.1. Lectins

Lectins are proteins, globulins, accumulated in the cotyledons' vacuoles, with at least one non-catalytic domain which bind reversibly to carbohydrates or glycoproteins [43].

Many lectins present in raw or under-cooked beans are resistant to acidic and enzymatic proteolysis being absorbed into the blood stream of the animals. The affinity of some lectins (phytohemagglutinin) to the red blood cells results in red blood cells agglutination and hemolytic anemia [38]. The levels of lectins are not influenced by the soaking process and cooking until getting soft beans (60 minutes) seems to be adequate to eliminate lectins' hemaglutinnating activity [44].

*In vitro* studies with the phytohemagglutinin (PHA) of *Phaseolus vulgaris* in cancer cell lines, such as SK-MEL-28, HT-144 and C32 human melanoma, showed the potential of *Phaseolus vulgaris'* lectin in inhibiting cancer cells [45]. *In vivo* studies with mice pre-treated with 0.2 g of PHA/kg, before starting oral 5-fluorouracyl (FU) revealed higher survival of intestinal epithelium functional cells than mice not pre-treated with lectin [46].

## 5.2. Phaseolin and small bioactive peptides

Phaseolin is a trimeric glycoprotein, highly resistant to *in vitro* and *in vivo* digestion, as a consequence of the compact structure given by the high percentage of β-strands, high glycosylation pattern and hydrophobicity. Heat treatment promotes structural changes in the tertiary and quaternary structures of the protein, increasing susceptibility to enzymatic proteolysis and digestibility [47]. Depending on the molecular weight of phaseolin subunits, phaseolin can be classified as S (Sanilac), T (Tendergreen) and I (Inca) [48].

The small peptides obtained from phaseolin hydrolysis have potential antioxidant and iron chelating activities. After hydrolysis, the phaseolin chelating activity increases highly, from 18%, before hydrolysis, to more than 81% after the hydrolytic treatment [49].

Besides the antioxidant activity, the common bean's bioactive peptides have also anti-hypertensive, through angiotensin-converting enzyme (ACE) inhibition, hypoglycemic, through $\alpha$-amylase, $\alpha$-glucosidase and dipeptidyl peptidase-IV (DPP-IV) inhibition and anti-carcinogenic properties, through cell apoptosis induction [50, 51].

## 5.3. Protease inhibitors

Serine protease inhibitors are traditionally divided into two families: the Kunitz trypsin inhibitors and the Bowman-Birk trypsin/chymotrypsin inhibitors. The Kunitz trypsin inhibitor is predominantly found in soybeans and the Bowman-Birk family is widely present in legume seeds. The Protease inhibitors of common bean (*Phaseolus vulgaris*) are included in the Bowman-Birk family [52]. Similar to the lectins, protease inhibitors protect plant from insects and predators and also protect the seed against fungi and microorganisms after harvesting, extending seeds' shelf life [53].

Protease inhibitors of raw or barely cooked legumes resist to the acidic pH of stomach and to the proteolytic enzymes (pepsin) and reach to the duodenum, interfering with digestion through irreversible inhibition of trypsin and chymotrypsin. Since, in duodenum, protease levels are reduced, protein digestibility is compromised and the absorption of amino acids decreases [54]. Despite the negative impact in serine proteases, the denatured protease inhibitors have several health-promoting benefits in human health, mostly as anti-inflammatory and anti-carcinogenic compounds in *in vitro* and *in vivo* models [55]. Until now the molecular mechanism underlying Bowman-Birk inhibition in colorectal chemoprevention remains unknown [56].

## 5.4. $\alpha$-Amylase inhibitors

The $\alpha$-amylase inhibitors are mostly found in the embryonic axes and cotyledons of the seed as a defensive strategy against predators. These inhibitors prevent starch digestion by blocking the active site of the $\alpha$-amylase enzyme [57]. The traditional cooking process at 100°C during 10 minutes inactivates $\alpha$-amylase inhibitors [57]. Several clinical studies with humans, conducted to characterize the effect of $\alpha$-amylase inhibitor from raw white beans in weight loss and blood glucose levels, clearly showed the potential of a concentrated extract of white bean, with 3000 $\alpha$-amylase inhibiting units per gram (before meals with carbohydrates) in reducing body weight, body mass index (BMI), fat mass, waist/hip circumferences, systolic/diastolic blood pressure, triglycerides and post-prandial spikes in blood sugar, maintaining the lean body mass [58, 59].

## 5.5. Phytosterols

Phytosterols include plant sterols and stanols. Plant sterols are the most predominant sterols in plants, corresponding to unsaturated compounds with a double bond in the sterol ring. $\beta$-sitosterol, campesterol and stigmasterol are examples of sterols. Stanols represent only 10%

of the total dietary phytosterols and are distinguished from sterols based on the absence of double bonds on the sterol ring (saturated molecules) [60].

Since humans cannot synthesize phytosterols, it must be achieved through the consumption of cereals, legumes, vegetables, fruits and nuts. In legumes, the sterols content is quite variable ranging from 134 mg/100 g, in kidney beans, to 242 mg/100 g, in peas [61]. Common bean show high levels of stigmasterol, 86.2 mg/100 g and 41.4 mg/100 g, in butter and kidney beans, respectively [61]. Dietary phytosterols intake normally ranges between 78 and 500 mg/day [62]. Some negative effects have been related to phytosterols consumption up to 1 year and include nausea, diarrhea or constipation. However, *in vivo* studies with rats associate phytosterols with several beneficial biological effects including anti-inflammatory and anticarcinogenic effects [60]. Phytosterols have been extensively studied as compounds with the ability to decrease cholesterol levels in the gut [42].

### 5.5.1. Phytates

Phytic acid is accumulated in plant seeds in the form of a salt associated with magnesium, calcium and copper, during the maturation stage. It represents 60–90% of the total phosphorus in the seed [63]. The phytate content in legumes is higher than in cereal-based food items. For instance, in cooked kidney beans it ranges from 8.3 to 13.4 mg/g dry weight (DW) while in wheat bread, the levels are considerably low (3.2–7.3 mg/g DW) [64].

Monogastric animals, including poultry and humans are unable to metabolize phytic acid as a consequence of the lack of the phytase degrading enzymes at gastrointestinal level [65].

The main anti-nutritional effects of phytates result from phytate capacity to chelate minerals such as calcium, zinc, copper and magnesium, reducing the minerals bioavailability on diet [66]. Phytates can also establish non-specific complexes with proteins which are less prone to digestion by proteolytic enzymes [67]. Processing strategies, such as soaking [68], germination [69], fermentation [64] and the addition of phytases in animal feed [70] and as food additives [71] promote dephosphorylation of phytate improving the nutritional value of legumes. Due to phytate heat-stability, cooking process does not affect phytate content [72]. Despite the anti-nutritional effects, phytates have been related to anti-oxidant effects [73], anti-carcinogenic activity [74], hypolipidemic [75] and hypoglicemic effects [76].

Regarding the impact of phytate in human health and the dose to ensure beneficial/negative effects, more studies are required and should be a priority for new research lines.

### 5.6. Saponins

Chemically referred as triterpene and steroid glycosides, saponins are formed by one or more carbohydrate units attached to a triterpenoid or steroidal aglycone (sapogenin) [77]. Saponins are soluble in water and its content is reduced during soaking process [78]. The lowest saponin content was obtained when beans were only soaked for 6 h [78]. Saponins can be responsible for a bitter taste and astringency that compromises food intake.

Recognized as anti-nutritional compounds, saponins may reduce nutrients' bioavailability and decrease trypsin and chymotrypsin activity [79]. Despite of the anti-nutritional effects, saponins have been explored as hypocholesterolemic [80] and hypoglycemic compounds [81]. Saponins have also been studied for their anticarcinogenic activity, considering in cell based assays with hepatocellular carcinoma cells (HepG2), fibrosarcoma cells (HT1080), cervical cancer cells (HeLa), promyelocytic leukemia cells (HL60) and breast cancer cells (MDA-MB-453) [82].

### 5.7. Phenolic compounds

Phenolic compounds, in common bean, include a huge diversity of secondary metabolites (phenolic acids such as hydroxybenzoic and hydroxycinnamic acids, flavonoids and stilbenes) synthesized from the amino acids phenylalanine or tyrosine, in the phenylpropanoid pathway. The C6-C1 skeleton of benzoic acids is generated by shortening of the hydroxycinnamic acids, **Figure 1**. Flavonoids are characterized by a C6-C3-C6 general structure, formed by two benzene rings (A and B) linked by a three carbon chain (a heterocyclic ring with an oxygen, the C ring), **Figure 1**. Stilbenes have a general structure C6-C2-C6, **Figure 1** [83].

Based on the chemical structure, flavonoids can be classified into six different classes, the flavones, flavanones, flavonols, flavanols, anthocyanins and isoflavones [84]. In **Table 1**, the major differences in the chemical structure of compounds included into the different flavonoids' classes are summarized [84].

In dry beans such as common bean, the majority of phenolic compounds are classified as phenolic acids and flavonoids (including proanthocyanidins). The anthocyanins, isoflavones, flavanols and flavonols are mostly located in the seed coat. The cotyledons are particularly rich in phenolic acids such as the hydroxycinnamic acids (e.g. ferulic and sinapic acids), mostly in esterified and glycosylated forms [85].

The content of phenolic compounds is quite variable depending on the legumes species, cultivar, seed's coat color pattern, maturity, growing location, environmental characteristics, storage conditions and processing techniques (e.g. boiling, germination and fermentation) [86]. The dark-colored varieties have higher qualitative and quantitative diversity of phenolic compounds, especially anthocyanins and proanthocyanidins, than lighter varieties [85]. Flavonols such as quercetin and kaempferol glycoside derivatives have been described in black, pinto, dark red kidney, light red kidney and small red beans collected in the USA [87], in Mexican black, mottled gray, caffeto and pale beans [88] and in the Italian yellow and black seed coat beans [89]. Nonglycosylated isoflavones (daidzein and genistein) have been identified by LC-ESI-QTOFMS in Brazilian black varieties of common bean [90]. The phenolic acids derived from benzoic and hydroxycinnamic acids have been studied in Mexican varieties of common beans [88]. The ferulic, sinapic, vanillic and p-hydroxybenzoic acids were the most abundant phenolic acids in the Mexican varieties, regardless of the seed coats' color [88].

Stilbene compounds such as resveratrol glucoside was identified and quantified, by mass spectrometry, in germinated black beans [91]. The anti-nutritional impact of phenolic compounds in human health is related to its inhibitory effect in the digestion enzymes (e.g. $\alpha$-amylase and pancreatic lipase) [92]. In legumes, particularly rich in tannins, phenolic compounds may also interact with dietary proteins, promoting proteins' precipitation or reducing protease (e.g. pepsin, trypsin and chymotrypsin) accessibility to the hydrophobic sites on the proteins, [93] which impairs protein digestibility.

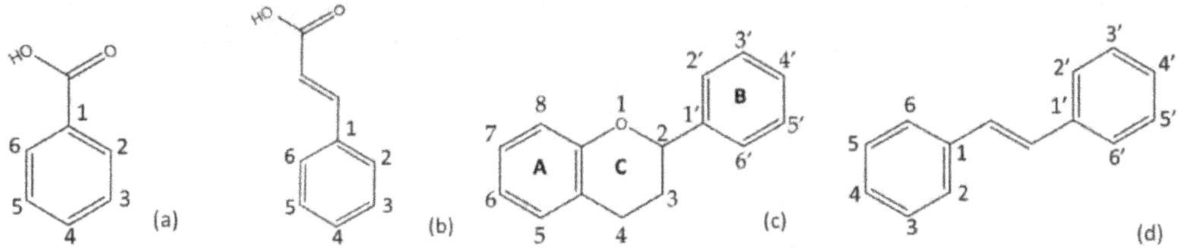

**Figure 1.** General structure of the most predominant phenolic compounds' families in common bean (a) *p*-hydroxybenzoic acids; (b) hydroxycinnamic acids; (c) flavonoids; (d) stilbenes.

| Flavonoids class | Chemical structure characteristics | Examples |
| --- | --- | --- |
| Flavones | • Double bond C2-C3 (unsaturated C ring) | Apigenin |
| | • Ketone at C4 of the C ring | |
| Flavanones (Dihydroflavones) | • Saturated C ring | Naringenin |
| | • Ketone at C4 of the C ring | |
| Flavonols | • Double bond C2-C3 (unsaturated C ring) | Quercetin |
| | • Ketone at C4 of the C ring | |
| | • OH- group at C3 of the C ring | |
| Flavanols (Flavan-3-ols or Catechins) | • Saturated C ring | Catechin |
| | • OH- group at C3 of the C ring | Procyanidin B1 |
| | • Ability to form polymers | |
| Anthocyanins | • Flavylium cations | Cyanidin |
| | • Majority in glycosidic form (sugars attached at C3) | Cyanidin-3-glucoside |
| Isoflavones | • Double bond C2-C3 (unsaturated C ring) | Daidzein |
| | • Ketone at C4 of the C ring | |
| | • B ring attached to C ring at C3 | |

**Table 1.** Chemical structure of compounds included into the different flavonoids' classes.

The negative impact of phenolic compounds in nutrients' digestibility can be glimpsed as a potential property of legumes to manage body weight and prevent obesity [93]. The health benefits of phenolic compounds are dependent on phenolic compounds' absorption and metabolism, which is influenced by several factors related to phenolic compounds' structure, molecular size, solubility, concentration in food, degree of glycosylation, phenolic compounds interaction and phenolic compounds matrix binding interaction, cell wall structure, as well as, by individual factors such as enzyme activity, intestinal transit time, genetics, gender, age, microflora composition and gastrointestinal pathologies [94]. The cluster of mentioned factors indicates that the most concentrated compounds are not necessarily the most bioavailable, in fact regardless of the abundance, most of the hydroxycinnamic acids are in the esterified form which compromises hydroxycinnamic acids' intestinal absorption and bioavailability [95].

The health-promoting effects of common bean phenolic compounds include the anti-oxidant [96], anti-inflammatory [97], anti-hyperglycemic [98], anti-hyperlipidemic [99] and anti-carcinogenic [100] activities. The molecular mechanisms responsible for such biological activities need further study. Moreover, long-term clinical studies are required to establish common beans bioactive compounds' benefits on human body.

## 6. Innovative food products

New "ready-to-eat" food products with inclusion of legumes as ingredients have been flooding the market. In the European Union, 3593 new products have been released between 2010 and 2014 [101].

In what regards common bean innovative food products, bean flour has been incorporated mostly in bakery products and snacks. In Mexico, whole wheat bread has been supplemented with 0.5% of freeze-dried black bean seed coat extract [102]. In Brazil, common bean flour has been added to rice flour and sugar in a proportion of 30:70:5%, respectively, to produce extruded breakfast flakes [103]. In Canada, a new bean snack, similar to pretzels, composed by 34% of navy bean flour has been developed. In North America, common bean flour has also been incorporated in other snacks such as potato chips and tortilla chips [104]. In Italy, biscuits have been prepared with wheat, maize and common bean flour at different proportions (26.7, 32.1, 50.0, 53.6 and 64.3% of bean flour). The biscuits prepared with a bean flour percentage of 26.7 and 32.1% were accepted with a score similar to the traditional biscuit [105].

The improved quality of the new food products that include legumes as ingredients represent a new market challenge and a concerted action between research community and food industry with divulgation of the potential health benefits should be mandatory to increase legumes consumption.

## 7. Conclusion(s)

The Fabaceae family includes a huge number of species that can bring diversity, nutrient supply and disease control to cropping systems. In Europe, beans production and consumption decreased drastically in the last decades. Nutritionally different from starchy foods, legumes have higher protein, amylose, fiber, folate and minerals contents. Therefore, the inclusion of legumes in a daily diversified diet is one of the best nutritional strategies to prevent malnutrition. Legumes are also a rich source of bioactive compounds (e.g. enzymatic inhibitors and phenolic compounds). The content of such compounds in plants is quite variable depending on the plant genotype and on the environmental and processing conditions. In fact, most of the anti-nutritional effects can be inactivated toward preparation and processing techniques (e.g. soaking, peeling, boiling, fermentation and germination). Recent research on the impact of bioactive compounds on health showed their potential to exert biological actions as anti-oxidant, anti-inflammatory, anti-hyperlipidemic, anti-hyperglycemic and anti-carcinogenic compounds.

Future research lines should focus on the characterization of legume genetic diversity, development of reliable and quick screening assays of quality-related traits to improve varieties in legume breeding programs, update of legume consumption in each country and bioavailability studies (including assays regarding the effective doses of bioactive compounds responsible for significant biological actions in clinical studies).

## Acknowledgements

The authors acknowledge the financial support provided by the FP7-EU project Strategies for Organic and Low-input Integrated Breeding and Management (SOLIBAM), FCT, Portugal for the funded project – "Exploiting Bean Genetics for food Quality and Attractiveness Innovation" (BEGEQA), PTDC/AGR-TEC/3555/2012BEGEQA and Elsa Mecha PhD fellowship (SFRH/BD/89287/2012), Maria Carlota Vaz Patto, FCT Investigator Program Development Grant (IF/01337/2014), and through R&D unit, UID/Multi/04551/2013 (GreenIT).

## Conflict of interest

The authors declare no conflict of interest.

## Author details

Elsa Mecha[1], Maria Eduardo Figueira[2], Maria Carlota Vaz Patto[1] and
Maria do Rosário Bronze[1,2,3]*

*Address all correspondence to: mbronze@ibet.pt

1 Instituto de Tecnologia Química e Biológica António Xavier, Universidade Nova de Lisboa, Oeiras, Portugal

2 iMED, Faculdade de Farmácia, Universidade de Lisboa, Av. das Forças Armadas, Lisboa, Portugal

3 Instituto de Biologia Experimental e Tecnológica, Oeiras, Portugal

## References

[1] Considine MJ, Siddique KHM, Foyer CH. Nature's pulse power: Legumes, food security and climate change. Journal of Experimental Botany. 2017;**68**(8):1815-1818. DOI: 10.1093/jxb/erx099

[2] Khokhar S, Owusu-Apenten RK. Antinutritional factors in food legumes and effects of processing. In: Owusu-Apenten RK, editor. The Role of Food, Agriculture, Forestry and Fisheries in Human Nutrition. 1st ed. Oxford: Encyclopedia of Life Support Systems (EOLSS); 2003. pp. 82-116

[3]  Bellucci E, Bitocchi E, Rau D, Rodriguez M, Biagetti E, Giardini A, et al. Genomics of origin, domestication and evolution of *Phaseolus vulgaris*. In: Tuberosa R, Graner A, Frison E, editors. Genomics of Plant Genetic Resources. Springer; Dordrecht. 2014. pp. 483-507. DOI: 10.1007/978-94-007-7572-5_20

[4]  Leitão ST, Dinis M, Veloso MM, Šatović Z, Vaz Patto MC. Establishing the bases for introducing the unexplored Portuguese common bean germplasm into the breeding world. Frontiers in Plant Science. 2017;8(1296). DOI: 10.3389/fpls.2017.01296

[5]  Mavromatis A, Arvanitoyannis I, Korkovelos A, Giakountis A, Chatzitheodorou VA, Goulas CK. Genetic diversity among common bean (*Phaseolus vulgaris* L.) Greek landraces and commercial cultivars: Nutritional components, RAPD and morphological markers. Spanish Journal of Agricultural Research. 2010;8:986-994. DOI: 10.5424/sjar/2010084-1245

[6]  FAOSTAT [Internet]. 2017. Available from: http://www.fao.org/faostat/en/#compare [Accessed: 27-03-2018]

[7]  Foyer CH, Lam H-M, Nguyen HT, Siddique KHM, Varshney RK, Colmer TD, et al. Neglecting legumes has compromised human health and sustainable food production. Nature Plants. 2016;2:16112. DOI: 10.1038/nplants.2016.112

[8]  FAO, IFAD, UNICEF, WFP, WHO. The State of Food Security and Nutrition in the World 2017. Building Resilience for Peace and Food Security 2017 [March 24, 2018]; Available from: http://www.fao.org/3/a-I7695e.pdf [Accessed: 27-03-2018]

[9]  Agency CID. Increasing Food Security CIDA's Food Security Strategy. 2010, Available from: http://www.international.gc.ca/development-developpement/assets/pdfs/partners-partenaires/key_partners-partenaires_cles/food-security-strategy-e.pdf [Accessed: 24-03-2018]

[10]  Drewnowski A. The cost of US foods as related to their nutritive value. The American Journal of Clinical Nutrition. 2010;92(5):1181-1188. DOI: 10.3945/ajcn.2010.29300

[11]  Messina V. Nutritional and health benefits of dried beans. The American Journal of Clinical Nutrition. 2014;100(suppl_1):437S-442S. DOI: 10.3945/ajcn.113.071472

[12]  Winham D, Webb D, Barr A. Beans and Good Health. 2008;45(3):201-209. Available from: http://admin.aghost.net/images/e0160001/ntodayoct08.pdf [Accessed: 03-04-2018]

[13]  USDA Food Composition Databases [Internet]. 2018. Available from: https://ndb.nal.usda.gov/ndb/search/list [Accessed: 28-03-2018]

[14]  Montoya CA, Lallès J-P, Beebe S, Leterme P. Phaseolin diversity as a possible strategy to improve the nutritional value of common beans (*Phaseolus vulgaris*). Foodservice Research International. 2010;43(2):443-449. DOI: 10.1016/j.foodres.2009.09.040

[15]  Temba MC, Njobeh PB, Adebo OA, Olugbile AO, Kayitesi E. The role of compositing cereals with legumes to alleviate protein energy malnutrition in Africa. International Journal of Food Science and Technology. 2016;51(3):543-554. DOI: 10.1111/ijfs.13035

[16] Nosworthy MG, House JD. Factors influencing the quality of dietary proteins: Implications for pulses. Cereal Chemistry. 2017;**94**(1):49-57. DOI: 10.1094/CCHEM-04-16-0104-FI

[17] Verma AK, Kumar S, Das M, Dwivedi PD. A comprehensive review of legume allergy. Clinical Reviews in Allergy and Immunology. 2013;**45**(1):30-46. DOI: 10.1007/s12016-012-8310-6

[18] Kasera R, Singh BP, Lavasa S, Prasad KN, Sahoo RC, Singh AB. Kidney bean: A major sensitizer among legumes in asthma and rhinitis patients from India. PLoS One. 2011;**6**(11):e27193. DOI: 10.1371/journal.pone.0027193

[19] Boye J, Zare F, Pletch A. Pulse proteins: Processing, characterization, functional properties and applications in food and feed. Food Research International. 2010;**43**(2):414-431. DOI: 10.1016/j.foodres.2009.09.003

[20] Boye JI. Food allergies in developing and emerging economies: Need for comprehensive data on prevalence rates. Clinical and Translational Allergy. 2012;**2**:25. DOI: 10.1186/2045-7022-2-25

[21] Sampson HA, O'Mahony L, Burks AW, Plaut M, Lack G, Akdis CA. Mechanisms of food allergy. Journal of Allergy and Clinical Immunology. 2018;**141**(1):11-19. DOI: 10.1016/j.jaci.2017.11.005

[22] Ibanez MD, Martinez M, Sanchez JJ, Fernandez-Caldas E. Legume: Cross-reactivity. Allergologia Imunopathologia. 2003;**31**:151-161

[23] Abeshu MA, Lelisa A, Geleta B. Complementary feeding: Review of recommendations, feeding practices, and adequacy of homemade complementary food preparations in developing countries—Lessons from Ethiopia. Frontiers in Nutrition. 2016;**3**:41. DOI: 10.3389/fnut.2016.00041

[24] Riascos JJ, Weissinger AK, Weissinger SM, Burks AW. Hypoallergenic legume crops and food allergy: Factors affecting feasibility and risk. Journal of Agricultural and Food Chemistry. 2010;**58**(1):20-27. DOI: 10.1021/jf902526y

[25] Hoover R, Zhou Y. *In vitro* and *in vivo* hydrolysis of legume starches by $\alpha$-amylase and resistant starch formation in legumes—A review. Carbohydrate Polymers. 2003;**54**(4):401-417. DOI: 10.1016/S0144-8617(03)00180-2

[26] Karri J, Parimalavalli R. Comparative study on chemical, functional and pasting properties of chickpea (non cereal) and wheat (cereal) starches. International Food Research Journal. 2015;**22**(2):677-683

[27] Dipnaik K, Kokare P. Ratio of amylose and amylopectin as indicators of glycaemic index and *in vitro* enzymatic hydrolysis of starches of long, medium and short grain rice. International Journal of Research in Medical Sciences. 2017;**5**(10):4502-4505. DOI: 10.18203/2320-6012.ijrms20174585

[28] Chawla R, Patil GR. Soluble dietary fiber. Comprehensive Reviews in Food Science and Food Safety. 2010;**9**(2):178-196. DOI: 10.1111/j.1541-4337.2009.00099.x

[29] Tan J, McKenzie C, Potamitis M, Thorburn AN, Mackay CR, Macia L. Chapter three—The role of short-chain fatty acids in health and disease. Advances in Immunology. 2014;**121**:91-119. DOI: 10.1016/B978-0-12-800100-4.00003-9

[30] Ohira H, Tsutsui W, Fujioka Y. Are short chain fatty acids in gut microbiota defensive players for inflammation and atherosclerosis? Journal of Atherosclerosis and Thrombosis. 2017;**24**(7):660-672. DOI: 10.5551/jat.RV17006

[31] Yadav BS, Sharma A, Yadav RB. Resistant starch content of conventionally boiled and pressure-cooked cereals, legumes and tubers. Journal of Food Science and Technology. 2010;**47**(1):84-88. DOI: 10.1007/s13197-010-0020-6

[32] Winham DM, Hutchins AM. Perceptions of flatulence from bean consumption among adults in 3 feeding studies. Nutrition Journal. 2011;**10**:128. DOI: 10.1186/1475-2891-10-128

[33] Schaafsma G, Slavin JL. Significance of inulin fructans in the human diet. Comprehensive Reviews in Food Science and Food Safety. 2015;**14**(1):37-47. DOI: 10.1111/1541-4337.12119

[34] Worku A, Sahu O. Significance of fermentation process on biochemical properties of *Phaseolus vulgaris* (red beans). Biotechnology Reports. 2017;**16**:5-11. DOI: 10.1016/j.btre.2017.09.001

[35] Bouchenak M, Lamri-Senhadji M. Nutritional quality of legumes, and their role in cardiometabolic risk prevention: A review. Journal of Medicinal Food. 2013;**16**(3):185-198. DOI: 10.1089/jmf.2011.0238

[36] Caprioli G, Giusti F, Ballini R, Sagratini G, Vila-Donat P, Vittori S, et al. Lipid nutritional value of legumes: Evaluation of different extraction methods and determination of fatty acid composition. Food Chemistry. 2016;**192**:965-971. DOI: 10.1016/j.foodchem.2015.07.102

[37] Russo GL. Dietary n−6 and n−3 polyunsaturated fatty acids: From biochemistry to clinical implications in cardiovascular prevention. Biochemical Pharmacology. 2009;**77**(6):937-946. DOI: 10.1016/j.bcp.2008.10.020

[38] Kouris-Blazos A, Belski R. Health benefits of legumes and pulses with a focus on Australian sweet lupins. Asia Pacific Journal of Clinical Nutrition. 2016;**25**(1):1-17

[39] Hunt JR. Bioavailability of iron, zinc, and other trace minerals from vegetarian diets. The American Journal of Clinical Nutrition. 2003;**78**(3):633S-639S. DOI: 10.1093/ajcn/78.3.633S

[40] Horimoto Y, Lim L-T. Effects of different proteases on iron absorption property of egg white hydrolysates. Foodservice Research International. 2017;**95**:108-116. DOI: 10.1016/j.foodres.2017.02.024

[41] Mulambu J, Andersson SM, Palenberg M, Pfeiffer W, Saltzman A, Birol E, et al. Iron beans in Rwanda: Crop development and delivery experience. African Journal of Food Agriculture Nutrition and Development. 2017;**17**(2):12026-12050. DOI: 10.18697/ajfand.78.HarvestPlus10

[42] Campos-Vega R, Loarca-Piña G, Oomah BD. Minor components of pulses and their potential impact on human health. Foodservice Research International. 2010;**43**(2): 461-482. DOI: 10.1016/j.foodres.2009.09.004

[43] Chrispeels MJ, Raikhel NV. Lectins, lectin genes, and their role in plant defense. The Plant Cell. 1991;**3**(1):1-9

[44] Thompson LU, Rea RL, Jenkins DJA. Effect of heat processing on hemagglutinin activity in red kidney beans. Journal of Food Science. 1983;**48**(1):235-236. DOI: 10.1111/j.1365-2621.1983.tb14831.x

[45] Loréa P, Goldschmidt D, Darro F, Salmon I, Bovin N, J Gabius H, et al. *In vitro* characterization of lectin-induced alterations on the proliferative activity of three human melanoma cell lines. Melanoma Research. 1997;**7**:353-363. DOI: 10.1097/00008390-199710000-00001

[46] Pusztai A, Bardocz S, W B Ewen S. Use of plant lectins in bioscence and biomedicine. Frontiers in Bioscience. 2008;**13**:1130-1140. DOI: 10.2741/2750

[47] Hayat I, Ahmad A, Masud T, Ahmed A, Bashir S. Nutritional and health perspectives of beans (*Phaseolus vulgaris* L.): An overview. Critical Reviews in Food Science and Nutrition. 2014;**54**(5):580-592. DOI: 10.1080/10408398.2011.596639

[48] De La Fuente M, López-Pedrouso M, Alonso J, Santalla M, Ron AMD, Álvarez G, et al. In-depth characterization of the phaseolin protein diversity of common bean (*Phaseolus vulgaris* L.) based on two-dimensional electrophoresis and mass spectrometry. Food Technology and Biotechnology. 2012;**50**(3):315-325

[49] Carrasco-Castilla J, Hernández-Álvarez AJ, Jiménez-Martínez C, Jacinto-Hernández C, Alaiz M, Girón-Calle J, et al. Antioxidant and metal chelating activities of *Phaseolus vulgaris* L. var. Jamapa protein isolates, phaseolin and lectin hydrolysates. Food Chemistry. 2012;**131**(4):1157-1164. DOI: 10.1016/j.foodchem.2011.09.084

[50] Mojica L, Chen K, Mejía EG. Impact of commercial precooking of common bean (*Phaseolus vulgaris*) on the generation of peptides, after pepsin–pancreatin hydrolysis, capable to inhibit dipeptidyl peptidase-IV. Journal of Food Science. 2015;**80**(1):H188-HH98. DOI: 10.1111/1750-3841.12726

[51] Heredia-Rodríguez L, de la Garza AL, Garza-Juarez AJ, Vazquez-Rodriguez JA. Nutraceutical properties of bioactive peptides in common bean (*Phaseolus vulgaris* L.). Journal of Food Nutrition and Dietetics. 2016;**2**(1):111

[52] Lajolo FM, Genovese MI. Nutritional significance of lectins and enzyme inhibitors from legumes. Journal of Agricultural and Food Chemistry. 2002;**50**(22):6592-6598. DOI: 10.1021/jf020191k

[53] Kim J-Y, Park S-C, Hwang I, Cheong H, Nah J-W, Hahm K-S, et al. Protease inhibitors from plants with antimicrobial activity. International Journal of Molecular Sciences. 2009;**10**(6):2860-2872. DOI: 10.3390/ijms10062860

[54] Guillamón E, Pedrosa MM, Burbano C, Cuadrado C, Sánchez MC, Muzquiz M. The trypsin inhibitors present in seed of different grain legume species and cultivar. Food Chemistry. 2008;**107**(1):68-74. DOI: 10.1016/j.foodchem.2007.07.029

[55] Clemente A, Sonnante G, Domoney C. Bowman-Birk inhibitors from legumes and human gastrointestinal health: Current status and perspectives. Current Protein & Peptide Science. 2011;**12**(5):358-373. DOI: 10.2174/138920311796391133

[56] Clemente A, Arques MC. Bowman-Birk inhibitors from legumes as colorectal chemopreventive agents. World Journal of Gastroenterology: WJG. 2014;**20**(30):10305-10315. DOI: 10.3748/wjg.v20.i30.10305

[57] Obiro WC, Zhang T, Jiang B. The nutraceutical role of the Phaseolus vulgaris α-amylase inhibitor. The British Journal of Nutrition. 2008;**100**(1):1-12. DOI: 10.1017/s0007114508 879135

[58] Celleno L, Tolaini MV, D'Amore A, Perricone NV, Preuss HG. A dietary supplement containing standardized *Phaseolus vulgaris* extract influences body composition of overweight men and women. International Journal of Medical Sciences. 2007;**4**(1):45-52

[59] Vinson JA, Kharrat HA, Shuta D. Investigation of an amylase inhibitor on human glucose absorption after starch consumption. The Open Nutraceuticals Journal. 2009;**2**: 88-91. DOI: 10.2174/1876396000902010088

[60] Ogbe RJ, Ochalefu DO, Mafulul SG, Olaniru OB. A review on dietary phytosterols: Their occurrence, metabolism and health benefits. Asian Journal of Plant Science and Research. 2015;**5**(4):10-21

[61] Ryan E, Galvin K, O'Connor TP, Maguire AR, O'Brien NM. Phytosterol, squalene, tocopherol content and fatty acid profile of selected seeds, grains, and legumes. Plant Foods for Human Nutrition. 2007;**62**(3):85-91. DOI: 10.1007/s11130-007-0046-8

[62] Racette SB, Spearie CA, Phillips KM, Lin X, Ma L, Ostlund RE Jr. Phytosterol-deficient and high-phytosterol diets developed for controlled feeding studies. Journal of the Academy of Nutrition and Dietetics. 2009;**109**(12):2043-2051. DOI: 10.1016/j.jada.2009.09.009

[63] Afinah S, Yazid AM, Anis Shobirin MH, Shuhaimi M. Phytase: Application in food industry. International Food Research Journal. 2010;**17**:13-21

[64] Greiner R, Konietzny U. Phytase for food application. Food Technology and Biotechnology. 2005;**44**(2):125-140

[65] Cowieson AJ, Bedford MR. The effect of phytase and carbohydrase on ileal amino acid digestibility in monogastric diets: Complimentary mode of action? World's Poultry Science Journal. 2009;**65**(4):609-624. DOI: 10.1017/s0043933909000427

[66] Bohn L, Meyer AS, Rasmussen SK. Phytate: Impact on environment and human nutrition. A challenge for molecular breeding. Journal of Zhejiang University. Science. B. 2008;**9**(3):165-191. DOI: 10.1631/jzus.B0710640

[67]  Woyengo TA, Nyachoti CM. Review: Anti-nutritional effects of phytic acid in diets for pigs and poultry—Current knowledge and directions for future research. Canadian Journal of Animal Science. 2013;93(1):9-21. DOI: 10.4141/cjas2012-017

[68]  Liang J, Han B-Z, Nout MJR, Hamer RJ. Effect of soaking and phytase treatment on phytic acid, calcium, iron and zinc in rice fractions. Food Chemistry. 2009;115(3):789-794. DOI: 10.1016/j.foodchem.2008.12.051

[69]  Azeke MA, Egielewa SJ, Eigbogbo MU, Ihimire IG. Effect of germination on the phytase activity, phytate and total phosphorus contents of rice (*Oryza sativa*), maize (*Zea mays*), millet (*Panicum miliaceum*), sorghum (*Sorghum bicolor*) and wheat (*Triticum aestivum*). Journal of Food Science and Technology. 2011;48(6):724-729. DOI: 10.1007/s13197-010-0186-y

[70]  Dersjant-Li Y, Awati A, Schulze H, Partridge G. Phytase in non-ruminant animal nutrition: A critical review on phytase activities in the gastrointestinal tract and influencing factors. Journal of the Science of Food and Agriculture. 2015;95(5):878-896. DOI: 10.1002/jsfa.6998

[71]  Matsuo A, Sato K. Utilization of Aspergillus niger phytase preparation for hydrolysis of phytate in foods. In: El-Samragy Y, editor. Food Additive. Rijeka: IntechOpen; 2012. pp. 125-138. DOI: 10.5772/32294

[72]  Karkle ENL, Beleia A. Effect of soaking and cooking on phytate concentration, minerals, and texture of food-type soybeans. Food Science and Technology. 2010;30:1056-1060

[73]  Sakač M, Čanadanović-Brunet J, Mišan A, Medić VĐ. Antioxidant activity of Phytic acid in lipid model system. Food Technology and Biotechnology. 2010;48(4):524-529

[74]  Verghese M, Rao DR, Chawan CB, Walker LT, Shackelford L. Anticarcinogenic effect of phytic acid (IP6): Apoptosis as a possible mechanism of action. LWT- Food Science and Technology. 2006;39(10):1093-1098. DOI: 10.1016/j.lwt.2005.07.012

[75]  Kang MY, Kim SM, Rico CW, Lee S-C. Hypolipidemic and antioxidative effects of rice bran and phytic acid in high fat-fed mice. Food Science and Biotechnology. 2012;21(1):123-128. DOI: 10.1007/s10068-012-0015-3

[76]  Kim SM, Rico CW, Lee SC, Kang MY. Modulatory effect of rice bran and phytic acid on glucose metabolism in high fat-fed C57BL/6N mice. Journal of Clinical Biochemistry and Nutrition. 2010;47(1):12-17. DOI: 10.3164/jcbn.09-124

[77]  Liener IE. Implications of antinutritional components in soybean foods. Critical Reviews in Food Science and Nutrition. 1994;34(1):31-67. DOI: 10.1080/10408399409527649

[78]  Shi J, Xue SJ, Ma Y, Li D, Kakuda Y, Lan Y. Kinetic study of saponins B stability in navy beans under different processing conditions. Journal of Food Engineering. 2009;93(1):59-65. DOI: 10.1016/j.jfoodeng.2008.12.035

[79]  Gemede HF, Ratta N. Antinutritional factors in plant foods: Potential health benefits and adverse effects. International Journal of Nutrition and Food Sciences. 2014;3(4):284-289. DOI: 10.11648/j.ijnfs.20140304.18

[80]  Shi J, Arunasalam K, Yeung D, Kakuda Y, Mittal G, Jiang Y. Saponins from edible legumes: Chemistry, processing, and health benefits. Journal of Medicinal Food. 2004;7(1):67-78. DOI: 10.1089/109662004322984734

[81]  Elekofehinti OO. Saponins: Anti-diabetic principles from medicinal plants-A review. Pathophysiology. 2015;22(2):95-103. DOI: 10.1016/j.pathophys.2015.02.001

[82]  Netala VR, Ghosh SB, Bobbu P, Anitha D, Vijaya T. Triterpenoid saponins: A review on biosynthesis, applications and mechanism of their action. International Journal of Pharmacy and Pharmaceutical Sciences. 2015;7(1):24-28

[83]  Dubrovina AS, Kiselev KV. Regulation of stilbene biosynthesis in plants. Planta. 2017;246(4):597-623. DOI: 10.1007/s00425-017-2730-8

[84]  Tsao R. Chemistry and biochemistry of dietary polyphenols. Nutrients. 2010;2(12): 1231-1246. DOI: 10.3390/nu2121231

[85]  Ranilla LG, Genovese MI, Lajolo FM. Polyphenols and antioxidant capacity of seed coat and cotyledon from Brazilian and Peruvian bean cultivars (Phaseolus vulgaris L.). Journal of Agricultural and Food Chemistry. 2007;55(1):90-98. DOI: 10.1021/jf062785j

[86]  Dueñas M, Sarmento T, Aguilera Y, Benitez V, Mollá E, Esteban RM, et al. Impact of cooking and germination on phenolic composition and dietary fibre fractions in dark beans (Phaseolus vulgaris L.) and lentils (Lens culinaris L.). LWT- Food Science and Technology. 2016;66:72-78. DOI: 10.1016/j.lwt.2015.10.025

[87]  Lin L-Z, Harnly JM, Pastor-Corrales MS, Luthria DL. The polyphenolic profiles of common bean (Phaseolus vulgaris L.). Food Chemistry. 2008;107(1):399-410. DOI: 10.1016/j. foodchem.2007.08.038

[88]  Espinosa-Alonso LG, Lygin A, Widholm JM, Valverde ME, Paredes-Lopez O. Polyphenols in wild and weedy Mexican common beans (Phaseolus vulgaris L.). Journal of Agricultural and Food Chemistry. 2006;54(12):4436-4444. DOI: 10.1021/jf060185e

[89]  Romani A, Vignolini P, Galardi C, Mulinacci N, Benedettelli S, Heimler D. Germplasm characterization of Zolfino landraces (Phaseolus vulgaris L.) by flavonoid content. Journal of Agricultural and Food Chemistry. 2004;52(12):3838-3842. DOI: 10.1021/jf0307402

[90]  de Lima PF, Colombo CA, Chiorato AF, Yamaguchi LF, Kato MJ, Carbonell SAM. Occurrence of Isoflavonoids in Brazilian common bean Germplasm (Phaseolus vulgaris L.). Journal of Agricultural and Food Chemistry. 2014;62(40):9699-9704. DOI: 10.1021/ jf5033312

[91]  López A, El-Naggar T, Dueñas M, Ortega T, Estrella I, Hernández T, et al. Effect of cooking and germination on phenolic composition and biological properties of dark beans (Phaseolus vulgaris L.). Food Chemistry. 2013;138(1):547-555. DOI: 10.1016/j. foodchem.2012.10.107

[92] Boath AS, Stewart D, McDougall GJ. Berry components inhibit α-glucosidase *in vitro*: Synergies between acarbose and polyphenols from black currant and rowanberry. Food Chemistry. 2012;**135**(3):929-936. DOI: 10.1016/j.foodchem.2012.06.065

[93] Velickovic TDC, Stanic-Vucinic DJ. The role of dietary phenolic compounds in protein digestion and processing technologies to improve their antinutritive properties. Comprehensive Reviews in Food Science and Food Safety. 2018;**17**(1):82-103. DOI: 10.1111/1541-4337.12320

[94] D'Archivio M, Filesi C, Varì R, Scazzocchio B, Masella R. Bioavailability of the polyphenols: Status and controversies. International Journal of Molecular Sciences. 2010;**11**(4):1321-1342. DOI: 10.3390/ijms11041321

[95] Ozcan T, Akpinar-Bayizit A, Yilmaz-Ersan L, Delikanli B. Phenolics in human health. International Journal of Chemical Engineering and Applications. 2014;**5**(5):393-396. DOI: 10.7763/IJCEA.2014.V5.416

[96] Oomah BD, Corbé A, Balasubramanian P. Antioxidant and anti-inflammatory activities of bean (*Phaseolus vulgaris* L.) hulls. Journal of Agricultural and Food Chemistry. 2010;**58**(14):8225-8230. DOI: 10.1021/jf1011193

[97] Zhang C, Monk JM, Lu JT, Zarepoor L, Wu W, Liu R, et al. Cooked navy and black bean diets improve biomarkers of colon health and reduce inflammation during colitis. British Journal of Nutrition. 2014;**111**(9):1549-1563. DOI: 10.1017/s0007114513004352

[98] Villegas R, Gao Y-T, Yang G, Li H-L, Elasy TA, Zheng W, et al. Legume and soy food intake and the incidence of type 2 diabetes in the shanghai women's health study. The American Journal of Clinical Nutrition. 2008;**87**(1):162-167

[99] Finley JW, Burrell JB, Reeves PG. Pinto bean consumption changes SCFA profiles in fecal fermentations, bacterial populations of the lower bowel, and lipid profiles in blood of humans. The Journal of Nutrition. 2007;**137**(11):2391-2398. DOI: 10.1093/jn/137.11.2391

[100] Campos-Vega R, García-Gasca T, Guevara-Gonzalez R, Ramos-Gomez M, Oomah BD, Loarca-Piña G. Human gut flora-fermented nondigestible fraction from cooked bean (*Phaseolus vulgaris* L.) modifies protein expression associated with apoptosis, cell cycle arrest, and proliferation in human adenocarcinoma colon cancer cells. Journal of Agricultural and Food Chemistry. 2012;**60**(51):12443-12450. DOI: 10.1021/jf303940r

[101] Agriculture and Agri-Food Canada. New Food Products with Pulse Ingredients Launched in the European Union. 2015. Available from: http://www.agr.gc.ca/resources/prod/Internet-Internet/MISB-DGSIM/ATS-SEA/PDF/6574-eng.pdf [Accessed: April 10, 2018]

[102] Chávez-Santoscoy RA, Lazo-Vélez MA, Serna-Sáldivar SO, Gutiérrez-Uribe JA. Delivery of flavonoids and saponins from black bean (*Phaseolus vulgaris*) seed coats incorporated into whole wheat bread. International Journal of Molecular Sciences. 2016;**17**(2):222. DOI: 10.3390/ijms17020222

[103]  Carvalho AV, Mattietto R de A, Bassinello PZ, Koakuzu SN, Rios A de O, Maciel R de A, et al. Processing and characterization of extruded breakfast meal formulated with broken rice and bean flour. Food Science and Technology. 2012;**32**:515-524

[104]  Anton AA, Luciano FB, Maskus H. Development of Globix: A new bean-based pretzel-like snack. Cereal Foods World. 2008;**53**(2):70-74. DOI: 10.1094/cfw-53-2-0070

[105]  Sparvoli F, Laureati M, Pilu R, Pagliarini E, Toschi I, Giuberti G, et al. Exploitation of common bean flours with low antinutrient content for making nutritionally enhanced biscuits. Frontiers in Plant Science. 2016;**7**:928. DOI: 10.3389/fpls.2016.00928

# Chapter 3

# Cowpea: A Strategic Legume Species for Food Security and Health

Alexandre Carneiro da Silva, Dyego da Costa Santos,
Davair Lopes Teixeira Junior, Pedro Bento da Silva,
Rosana Cavalcante dos Santos and Amauri Siviero

Additional information is available at the end of the chapter

http://dx.doi.org/10.5772/intechopen.79006

## Abstract

In this chapter, several characteristics of cowpea (*Vigna unguiculata*), including nutritional and nutraceutical properties, and economic and social aspects of production were analysed with the objective to demonstrate that cowpea is a culture suitable for inclusion in food security programs. Cowpea is rich in diverse nutrients, highlighting high levels of protein. Cowpea also is rich in nutraceuticals compounds such as dietary fibre, antioxidants and polyunsaturated fatty acids and polyphenols. Widely cultivated and consumed cowpea is the very important legume for the nutrition and health of millions of people in many countries. In addition to being nutritious and safe, cowpea has high relative productivity, production stability and high tolerance to environmental stresses such as drought. Cowpea also has economic viability, low environmental impact and contributes to the conservation of natural resources and the sustainability of production systems. Cowpea is a safe food, always available in most regions, low priced compared to other sources of protein. Based on the analyses performed, it is possible to infer that cowpea is a strategic culture for the promotion of food security and health of populations on all continents.

**Keywords:** diseases prevention, nutraceuticals, nutrition, phytochemicals, *Vigna unguiculata*

## 1. Introduction

At the historic reunion of 1996 in Rome, in World Food Summit of Food and Agriculture Organization of the United Nations (FAO), food security is met when "all people, at all times have physical and economic access to sufficient, safe and nutritious food that meets their

dietary needs and food preferences for an active and healthy life". Due to its central role in human development, food security is recognized as a universal human right [1].

Promoting food security is a complex mission with political, economic, environmental, social and cultural dimensions. For food security to be achieved, the population should have unrestricted access to a healthy and nutritious diet, which depends on adequate economic resources and food available in the country, region and communities in which people are located. The national availability of food for human consumption is a function of the balance between food grown in the country, import and export of food, reduction of waste and destination of food [2]. At the cultural and sociopolitical level, countries must provide incentives for advancing locally based and culturally relevant ethnic foods. The critical knowledge and creativity gained from long-term accumulated traditional knowledge can inform contemporary food science and nutritional and health science to advance more sustainable strategies based on experiences of diverse ecologies and cultures around the world [3].

The health is directly related with the balanced nutrition. Hippocrates "the father of medicine", over two millennia ago, mentioned about 400 medicinal plants and uttered the maxim, "let food be your medicine and let medicine be your food" [4]. Nutraceutical is the hybrid of 'nutrition' and 'pharmaceutical'. Nutraceuticals, in broad, are food or part of food playing a significant role in modifying and maintaining normal physiological function that maintains healthy human beings [5]. Grain legumes contain numerous phytochemicals useful for their nutritional or nutraceutical properties [6].

During the last decade, legumes have emerged as an interesting and balanced source of nutrients, being currently widely cultivated and consumed in different parts of the world [7, 8]. Legumes are consumed worldwide as an alternative source of proteins, since they are rich in amino acids like lysine and tryptophan and they are much cheaper than animal proteins [5]. Legumes are an excellent source of many essential nutrients, including vitamins, minerals, fibres, antioxidants and other bioactive compounds [9–11], including enzyme inhibitors, lectins, phytates, oligosaccharides and phenolic compounds that play metabolic roles in humans consuming these foods frequently [12].

The health organizations around the world recommend consuming legumes as part of a healthy diet, particularly because they have an important role in the control and prevention of chronic non-communicable diseases (NCDs) such as diabetes, cardiovascular diseases and cancer [13–15]. The legumes also favour the control of body weight, since they give greater satiety, prevent the accumulation of fat at the abdominal level and regulate blood sugar levels [15–17].

Cowpea (*Vigna unguiculata* [L.] Walp) is grain legume originated in the African continent with large economic and social importance in the developing world. Cowpea is a food of major importance for millions of people, especially in less developed countries of the tropics, being the major source of protein and carbohydrate dietary of the large part of the world population. Cowpea is not only rich in nutrients, but also nutraceuticals such as dietary fibre, antioxidants and polyunsaturated fatty acids (PUFA) and polyphenols [18–21].

Widely consumed in many countries, with excellent nutritional and nutraceutical properties and several agronomic, environmental and economic advantages, contributing to food security and maintenance of environment [22, 23]; cowpea is a strategic culture for the promotion of food security and health of populations on all continents.

## 2. Production and food security

### 2.1. Aspects of production

Cowpea is one of the key food sources in the arid, semi-arid and tropical parts of Asia, Oceania, southern Europe, Africa, southern United States and Central and South America [24]. Cowpea is grown as a main legume crop in Africa (Egypt, Nigeria), South America (Colombia, Brazil), the USA, Mexico, Asia (China, Pakistan and Japan) and in South and Southeast Europe (Spain, Italy, Portugal, Greece and Cyprus) [25]. It is truly a multifunctional crop, providing food for man and livestock and serving as a valuable and dependable revenue-generating commodity for farmers and grain traders [24, 26–27].

According to the data from the Food and Agriculture Organization (FAO) (http://www.fao.org), approximately 5.8 million tons of dry cowpea cereal is produced annually with a minimum of 11 million hectares planted all over the world [27], an average productivity of 527 kg ha$^{-1}$. However, due to the low productivity in regions that grow cowpea for subsistence, with low technological level and use of traditional genotypes, this productivity is below the potential of the crop that is 6000 kg ha$^{-1}$ [28].

Recent studies to evaluate the "adaptability and stability" and the productive performance of different genotypes (e.g. cultivar, lineage and hybrid) in different environmental conditions has allowed to obtain dry grain yields close and/or higher than 3000 kg ha$^{-1}$ for various genotypes [29, 30]. In addition to the positioning of genotypes suitable to crop environments, simple adjustments in production systems such as determination of planting time, spacing and plant density, are capable of promoting large productivity increases. Therefore, increased productivity and economic viability of cowpea is possible in all growing regions, using appropriate genotypes and improvements in production systems, reducing dependence on external inputs.

Cowpea is an essential component of sustainable cropping systems in the sub-humid tropics and, generally, dry regions across the globe. Cowpea is particularly important as a rotation crop with cereals. Cowpea can enhance the fertility of the soil with respect to nitrogen and phosphate, thereby benefiting subsequent cereal crops [23, 24]. Cowpea has great realization capacity of "biological nitrogen fixation (BNF)". BNF converts the atmospheric dinitrogen (N2) into usable nitrogen (N) by plants. BNF occurs in specialized plant structure called nodules formed by the symbiosis between roots and diazotrophic bacteria, which confer to leguminous crops the ability to satisfy their own and other plants' N-source demand [31, 32].

Cowpea is able to fix N in an amount greater than 100 kg ha$^{-1}$, replacing nitrogen fertilization [33], contributes to the low production cost of this culture [34]. BNF has contributed to the increase in cowpea yield, which along with other technological strategies has led to the expansion of the culture to news agricultural frontiers, competing as off-season culture with traditional commodities, such as corn [35]. Besides the fixation of N, the inclusion of cowpea in the rotation crop systems favours the accumulation of organic matter and greater fixation of carbon. This accumulation of organic matter contributes to the improvement of soil fertility and physical characteristics such as water infiltration and retention capacity, soil conservation and sustainability of production systems.

## 2.2. A strategic culture for food security

Increasing demands for nutritious, safe and healthy food because of a growing population and the pledge to maintain biodiversity and other resources pose a major challenge to agriculture that is already threatened by changing climate [36]. The access to healthy diet depends on the availability of nutritious foods at prices compatible with the purchasing power of populations [2]. Therefore, for a crop to be included in food safety programs, in addition to being nutritious and safe, it must have high relative productivity, production stability and high tolerance to environmental stresses (e.g. drought, salt soil, high temperature). They must also have economic viability, low environmental impact and contribute to the conservation of natural resources and the sustainability of production systems.

Cowpea is one of the most important edible grain legumes in underdeveloped and developing countries contributing to food security and maintenance of environment for millions of small-scale farmers and of the local populations [22, 37–38]. In developed countries, cowpea is also considered as a healthy alternative to soya bean as consumers look to more traditional food sources that are low in fat and high in fibre and that have other health benefits [39].

The availability of food is directly related to the agricultural production policies of each nation, which defines the agricultural crops that will receive investments in research, development of production technologies, financing, as well as the destination of production. This is especially relevant in case of ethnically and culturally relevant legumes, such as cowpea, where food support and subsidies in many countries favour a restricted choice of cereal crops over balanced co-production of legumes [40].

The research investment combined with the wide genetic diversity allowed us to obtain high-yield productive cowpea genotypes, early maturity and with plant architecture favourable to the mechanized harvest. These genotypes have greater resistance to the adverse environmental conditions to the cultivar (e.g. dry and temperature variation) and the attack of pests and diseases. In the last decade, cowpeas have ceased to be a subsistence crop, cultivated largely by family farmers and have aroused the interest of large farmers. With this, the relative increase in cowpea production in the first decade of the twenty-first century surpassed all other pulses[1] [41]. Cowpea is truly a multifunctional crop strategic for food security.

---

[1] Pulses are defined by the FAO as "limited to crops harvested solely for dry grain, thereby. Pulses exclude vegetable crops such as green peas and green beans, crops which are used primarily such oil crops (e.g. soybeans) and leguminous forage crops, such as alfalfa [42].

# 3. Nutritional properties

Cowpea plays a critical role in the lives of millions of people in the developing world, providing them a major source of dietary protein that nutritionally complements low-protein cereal and tuber crop staples [39]. With recognised nutritional value, cowpea can be consumed such as mature beans (i.e. dried grain), green beans or green pods. The cowpea leaves also can be consumed as food. Grains, pods and leaves of the cowpea are processed and used as food ingredient by the food industry [43–46].

The cowpea seeds are eaten boiled, parched, fried, roasted, mixed with sauce or stewed and consumed directly. Its seeds provide important vitamins, phytonutrients including antioxidants besides carbohydrates, minerals and trace elements. Cowpea due to its nutrients and functional benefits has also gained industrial importance for being used as a potential ingredient in food formulations [47].

Regarding the need for consumption, nutrients in the human diet can be classified as macronutrients (primary contributors to energy intake, which include total carbohydrate, total fat, protein and alcohol), micronutrients (minerals, vitamins and dietary fibre), and to include other food components such as bioactive compounds [48].

The consumption of cowpea supplies most of the macro and micronutrients of the diet. Chemical composition and nutritional properties of cowpeas vary considerably according to cultivar. For effective utilization of newly developed cowpea cultivars for human nutrition, the removal or reduction of antinutrients and evaluation of their nutritional properties are necessary [49].

## 3.1. Macronutrients

### 3.1.1. Protein

The nutritional profile of cowpea grain is similar to that of other pulses with a relatively low fat content and total protein content that is two- to fourfold higher than cereal and tuber crops [39]. Under the Harvest Plus initiative funded by the Bill & Melinda Gates Foundation and others, a systematic breeding program to develop improved cowpea varieties with enhanced levels of protein and micronutrient contents was initiated in 2003. Approximately, 2000 genotypes (e.g. cultivars and breeding lines) have been evaluated revealing significant genetic variability in seed protein contents, with values ranging from 21 to 30.7% [39]. The nutritional ranking of 30 Brazilian genotypes of cowpea revel protein contents ranging from 17.4 to 28.3%. [50]. For improved cowpea breeding lines, the protein content can be bigger than 30% [51, 52].

Similar to other pulses, the storage proteins in cowpea seeds are rich in the amino acids lysine and tryptophan when compared to cereal grains, but low in methionine and cysteine when compared to animal proteins [39]. Cowpea possess some undesirable properties that are common to other legume seeds, such as methionine and cysteine deficiency as well as considerable contents of antinutritional factors like protease inhibitors, lectins, phytic acid, tannins, among others [49, 53].

A protein has a good amino profile when it presents all the essential amino acids (i.e. those that cannot be synthesised by the body and therefore must be obtained by the diet) in significant quantity [54]. Genetic and agronomic factors may influence the amino acid profile of cowpea [55]. Analyses carried out by Frota et al. [56] and Vasconcelos et al. [57] have shown that cowpea presented cysteine and methionine as limiting amino acids, whereas the other essential amino acids met the recommendations of the amino acid standard of the FAO/WHO [58] for children (2–5 years). However, other authors have found values of all the essential amino acids below the recommendation in some cowpea cultivars [59, 60].

Given its nutritional value as well as the reduced environmental impact of the production systems, intense research efforts must be redirected to the evaluation of nutrients and antinutrients of cowpea, its digestibility and development of processing alternatives that may allow the production of foods with lower impact on human health as well as its potential contribution to human nutrition [7]. Conventional processing methods, such as soaking, boiling, germination and fermentation, are widely used to decrease the content of these undesirable components, which results in enhanced acceptability and nutritional quality in addition to optimal utilisation of this legume as human food [61].

Cowpea protein isolate is an alternative for incorporation into food products [56]. Protein isolation is an alternative for the minimisation of antinutritional factors, improved digestibility and bioavailability of leguminous amino acids [62]. A mixed food of legumes and cereals, particularly in developing countries, can compensate deficiencies or a low level of lysine and sulphur amino acids, in cereals and grain legumes, respectively [63]. The utilization of a nutritional quality index will allow pinpointing the genotypes that gather the largest number of desirable nutritional attributes and then assist in the planning of new crosses in the breeding program [50].

### 3.1.2. Carbohydrate

Cowpea is one of the main sources of calories for a large segment of world population [18, 57]. Cowpea seeds contain approximately 53–66% carbohydrate, most of which is found in the form of starch, has high amylose content and C-type starch crystallinity [64–66].

Legumes, such as cowpea, contain a considerable amount of resistant starch (i.e. starch that resists to digestion by amylase in the small intestine and progresses to the large intestine for fermentation by the gut bacteria) and also have a higher ratio of slow-digestible to rapid-digestible starch, compared to other carbohydrate foods. Resistant starch is associated with reduced glycemic response, which can be beneficial to insulin-resistant individuals and those with diabetes [67–69]. Carbohydrates that are digested slowly also result in a low glycemic index (GI) [70]. The consumption of low GI foods could prevent the emergence of several diseases, such as obesity, diabetes, cardiovascular diseases and even certain cancers [71]. Other important constituents in cowpea seeds are the $\alpha$-galactosides, with a recognized prebiotic function [72].

### 3.1.3. Lipids

Recently, cowpea has been stressed on a low fat content, comparatively to other legumes (chickpea, split pea, lentil, green gram and lupine), which makes it, according to nutritional guidelines, a legume with potential application in weight restriction diets [7]. The content and profile of lipids in cowpea seeds, such as other nutrients, vary among genotypes and are also influenced by environmental conditions during cultivation. According to Brazilian Agricultural Research Corporation (EMBRAPA), the content of lipids in cowpea seeds, on average, is 2% [73]. The nutritional ranking of 30 Brazilian genotypes of cowpea revel lipids contents range from 1.0 to 1.6% [50]. Frota et al. [74], found lipid content of 2.2% in seed cowpea BRS-Milênio, a cultivar obtained by genetic improvement, and its fatty acids profile was 29.4% saturated and 70.7% unsaturated. Iqbal et al. [13], on the other hand, obtained approximately double the lipid content (4.8 g 100 g$^{-1}$) in relation to the cultivar BRS-Milênio analysed in the study performed by Frota et al. [74].

The triglycerides are the most abundant lipids in the cowpea seeds, corresponding to 41.2% of the total fat. The cowpea seeds lipid profile includes still 25.1% of phospholipids, 10.6% of monoglycerides, 7.9% of free fatty acids, 7.8% of diglycerides, 5.5% of sterols and 2.6% of hydrocarbons + sterol esters [75, 76]. Of the total fatty acids, most of it (40.1–78.3%) consists of polyunsaturated fatty acids. Ranging from 20.5 to 67.1%, the palmitic acid is the most abundant fatty acid. The content of linoleic acid can be ranging from 20.8 to 40.3% and of the linolenic acid ranges from 9.6 to 30.9%. In smaller proportion (2.9–14.0%), the stearic acids complete the profile of fatty acids of the cowpea seeds [7].

### 3.2. Micronutrients

Micronutrients are organic or inorganic compounds present in small amounts and are not used for energy, but are nonetheless needed for good health. Nonessential micronutrients encompass a vast group of unique organic phytochemicals that are not strictly required in the diet, but when present at sufficient levels are linked to the promotion of good health. Essential micronutrients in the human diet include 17 minerals and 13 vitamins required at minimum levels to alleviate nutritional disorders (See [77]).

*3.2.1. Vitamin*

Cowpea is rich in vitamin A and C and also has appreciable amount of thiamin, riboflavin, niacin, vitamin B6 and pantothenic acid as well as small amount of foliate [78]. Vitamins are indispensable to the maintenance of various functions of physiological importance such as muscle contractility, nerve function, blood coagulation, digestive processes and acid–base balance [79].

The major vitamins present in cowpea are those belonging to the B complex, being reported in the following decreasing order: niacin (7.0–40.0 × 10$^{-3}$ g kg$^{-1}$) > panthothenic acid (17.0–22.0 × 10$^{-3}$ g kg$^{-1}$) > thiamine (2.0–17.0 × 10$^{-3}$ g kg$^{-1}$) > pyridoxine (2.0–4.0 × 10$^{-3}$ g kg$^{-1}$) > folic acid (1.0–4.0 × 10$^{-3}$ g kg$^{-1}$) > riboflavin (1.0–3.0 × 10$^{-3}$ g kg$^{-1}$) > biotin(0.2–0.3 × 10$^{-3}$ g kg$^{-1}$) > cobalamin (traces). Cowpea appears to be a particularly good source of vitamin C, with levels in seeds ranging from 52.0 to 554.0 × 10$^{-3}$ g kg$^{-1}$. Carotenoids, precursors of vitamin A, are also present in cowpea contributing to the antioxidant compounds provided by this legume. Lastly, from the various vitamin E vitamers present in cowpea, δ-tocopherol has been observed with the highest concentration (15.1–109.7 × 10$^{-3}$ g kg$^{-1}$), followed by γ-tocopherol (4.3–92.3 × 10$^{-3}$ g kg$^{-1}$), and γ-tocotrienol (0.7–3.4 × 10$^{-3}$ g kg$^{-1}$). The vitamin E composition of cowpea seems to differ significantly from that of most legumes, where γ-tocopherol dominates (reviewed by [7]).

*3.2.2. Minerals*

An appropriate intake of micro minerals is necessary for the human organism to meet its met-abolic needs, and hence avoid a wide range of associated health problems [80, 81]. Cowpea is rich in potassium with good amount of calcium, magnesium and phosphorus. It also has small amount of iron, sodium, zinc, copper, manganese and selenium [78].

The mineral composition of 30 newly developed Brazilian cowpea genotypes obtained by conventional plant breeding reveals the following contents of minerals in the seeds: iron 61–81 ppm; zinc 27–44 ppm; sodium 84–177 ppm; potassium 9570–12,510 ppm; calcium 290–440 ppm; magnesium 1310–1160 ppm; manganese 17–29 ppm; copper 20–22 ppm [50].

The cowpea seed analysis of 87 lines originated from a set of crosses involving 3 accessions of the IITA ('IT97K-1042-3', 'IT99K-216-48-1' and 'IT97K-499') and accessions adapted for cultivation in the Brazilian semi-arid tropical areas ('BRS Tapaihum', 'BRS Pujante' and 'Canapu'), reveals high variation in minerals values: calcium 420–6260 ppm; iron 42.0–137.0 ppm; zinc 38.0–55.5 ppm; potassium 21,000–27,000 ppm; sodium 29.2–88.0 ppm [80]. In the analysis of approximately 2000 cowpea genotypes under the Harvest Plus initiative [39], the typical calcium, iron, zinc and potassium values showed large variations among genotypes (e.g. calcium ranging from 545 to 1300 ppm and zinc ranging from 23 to 48 ppm). These analyses reveal that cowpea have high levels of these micronutrients in comparison with other cultures, and that the variation in the levels of these micronutrients favours the genetic improvement of the species and the obtaining of more nutritious genotypes [39, 80].

# 4. Nutraceuticals compounds

Nutraceuticals, in broad, are food or part of food playing a significant role in modifying and maintaining normal physiological function that maintains healthy human beings [5]. The food sources used as nutraceuticals are all natural and can be categorized as dietary fibre, probiotics, prebiotics, polyunsaturated fatty acids, antioxidant vitamins, polyphenols and other different types of herbal/natural foods [5, 82, 83].

## 4.1. Dietary fibre

Dietary fibre has been shown to have important health implications in the prevention of risks of chronic diseases such as cancer, CVD and diabetes mellitus [84]. Dietary fibre can be soluble or insoluble depending upon solubility in water. Cowpea has high level of dietary fibre, mainly of insoluble fibre. Water-soluble fibre can form viscous solutions. The insoluble fibre (i.e. lignin, cellulose and hemicellulose) has high water-holding capacity and acts on the regulation of the defecation process [12, 13], while that the soluble fibre can contribute for reducing the postprandial blood glucose and insulin levels, and serum cholesterol [85].

In cowpea flours, (i.e. dehulled, ground and defatted cotyledons) the total dietary fibre (calcu-lated as % of dry matter) is 14.1 ± 0.3, being 1.0 ± 0.0% of soluble fibre and 13 ± 0.2 of insoluble fibre

[86]. The dietary fibre composition of 30 newly developed Brazilian cowpea genotypes showed great genetic variability among the genotypes, with values ranging from 19.5 to 35.6 g 100 g$^{-1}$ [50].

## 4.2. Probiotics

Probiotics can be defined as live microbial feed supplements that beneficially affect the host animal by improving its intestinal microbial balance [87]. Food cultures that have such beneficial effects on human health have been termed "probiotic" [88].

Probiotics are associated with fermented foods, latter having a long tradition of acceptability in communities where they are produced, safe use and the established as well as postulated claims of health benefits [89, 90]. The probiotic potential of fermented foods, such as cowpea, sorghum and peanut plant seed extracts have been reported [91].

## 4.3. Prebiotics

Prebiotics are non-digestible food ingredients that selectively stimulating the growth and/or activity of one or a limited number of beneficial bacteria in the colon [92, 93]. The prebiotics resist hydrolysis by digestive enzymes and/or are not absorbed in the upper part of the gastrointestinal tract and pass into the large bowel and promote the growth of *Bifidobacterium* and *Lactobacillus*, contributing for the right balance of intestinal bacterial flora and the immune system. The growth of *Bifidobacterium* and *Lactobacillus* to dominate pathogenic organisms and thus invigorate human health is facilitated by certain oligosaccharides [94, 95]. Cowpea seeds are rich in $\alpha$-galactosides (raffinose, stachyose and verbascose) [7, 96], also known as the raffinose family oligosaccharides (RFOs) [99].

The $\alpha$-galactosides are beneficial compounds when ingested in amounts up to 3 g day$^{-1}$. However, when consumed in high doses the $\alpha$-galactosides can cause flatulence and interference with the absorption of other nutrients during the digestive process [97]. As RFO act as substrate for intestinal bacteria, they are also considered as prebiotics [98]. The Galactosyl-cyclitols, present in legume seeds, are considered as important phytochemicals related to disease prevention [99].

## 4.4. Polyunsaturated fatty acids

Polyunsaturated fatty acids (PUFAs) are also called "essential fatty acids" as these are crucial to the body's function and are introduced externally through the diet [100]. PUFAs have two subdivisions: omega-3- (n-3) fatty acids and omega-6-(n-6) fatty acids. In cowpea, the bulk of fatty acids consist of polyunsaturated fatty acids that range from 40.1 to 78.3% of total (reviewed by [7]). This high level of unsaturated fatty acids is a nutritionally desirable feature [101].

Studies suggest that PUFAs have therapeutic effects in cardiovascular and hypolipidemic diseases. Emerging research evidence shows the benefits of omega-3-oils in other areas of health including premature infant health, asthma, bipolar and depressive disorders, dysmenorrhea and diabetes (reviewed by [5]).

### 4.5. Antioxidant vitamin

Cowpea appears to be a particularly good source of vitamin C, with levels in seeds and pods ranging from 52.0 to 554.0 × 10$^{-3}$ g kg$^{-1}$ [7]. Carotenoids, precursors of vitamin A, are also present in cowpea contributing to the antioxidant compounds provided by this legume. Among the carotenoids present in cowpea seeds, lutein makes up over 70.0%.Other carotenoids present in cowpea are β-carotene, γ-carotene and cryptoxanthin [7, 102].

From the various vitamin E vitamers present in cowpea, δ-tocopherol has been observed with the highest concentration, followed by γ-tocopherol -tocotrienol (0.7–3.4 × 10$^{-3}$ g kg$^{-1)}$ [7]. The vitamin E composition of cowpea seems to differ significantly from that of most legumes, where γ-tocopherol dominates [103, 104].

### 4.6. Phenolic compounds

Cowpea is a good source of dietary phenolics mainly phenolic acids, flavonoids and anthocyanins and proanthocyanidins. These compounds are reportedly responsible for the antioxidant and other health promoting properties of cowpea [105].

Phenolic compounds (tannins, flavonoids and phenolic acids) are secondary metabolites in plants and are present in some plant foods [106, 107]. Phenolic compounds are responsible for various beneficial effects in a multitude of diseases [108]. Phenolic compounds have antioxidant properties and ability to modulate the activity of various enzymes. These phenolics are also potent inhibitors of a-amylase and a-glucosidase, the two important enzymes involved in the regulation of glucose homeostasis [109].

## 5. Conclusion

In the current scenario of population growth, the demand for nutritious and functional, safe, and healthy food poses a major challenge for producers, which in times of climate change are enjoined to conserve natural resources, and for the governments of nations in need invest in the production of crops that can be included in food security programs and contribute to the health of populations.

With excellent nutritional and nutraceutical properties and several agronomic, environmental and economic advantages, cowpea is able of contribute to food security, maintenance of environment and promotion health for populations. This is possible due to the great genetic variability of the cowpea and the numerous researches to develop new genotypes more productive, biofortified, adapted to different environments and production systems. Cowpea is a much studied culture, which has its composition known, with great value for the food industry.

The use of cowpea as functional food has encouraged the industry and farmers to produce this legume. Considering the great demand worldwide for consumption, the excellent nutritional and nutraceutical properties, the availability of production technology and the wide possibility of choice of genotypes for production, cowpea is undoubtedly a strategic legume specie for food security and health.

# Acknowledgements

The author thank the Coordination for the Improvement of Higher Education Personnel (CAPES-Brazil).

# Author details

Alexandre Carneiro da Silva[1]*, Dyego da Costa Santos[1], Davair Lopes Teixeira Junior[1], Pedro Bento da Silva[2], Rosana Cavalcante dos Santos[3] and Amauri Siviero[4]

Address all correspondence to: alexandre.silva@ifac.edu.br

1 Federal Institute of Education, Science and Technology of Acre (IFAC), Xapuri, Acre State, Brazil

2 Sacred Heart University (USC), Bauru, São Paulo State, Brazil

3 Federal Institute of Education, Science and Technology of Acre (IFAC), Rio Branco, Acre State, Brazil

4 Brazilian Agricultural Research Corporation, Rio Branco, Acre State, Brazil

# References

[1] FAO. World Food Summit [Internet]. Rome (Italy): FAO; 1996. [cited 2018 Abr 5]. Available from: http://www.fao.org/wfs/index_en.htm

[2] Pérez-Escamilla R. Food Security and the 2015-2030 Sustainable Development Goals: From Human to Planetary Health: Perspectives and Opinions. Current Developments in Nutrition. 2017;1(7):e000513

[3] Shetty K, Sarkar D. Advancing Ethnic Foods in Diverse Global Ecologies through Systems-Based Solutions is Essential to Global Food Security and Climate Resilience-Integrated Human Health Benefits; 2018

[4] Shultes RE. The kingdom of plants. In: Thomson WAR, editor. Medicines from the Earth. New York, NY: McGraw-Hill Book Co; 1978. p. 208

[5] Das L, Bhaumik E, Raychaudhuri U, Chakraborty R. Role of nutraceuticals in human health. Journal of Food Science and Technology. 2012;49(2):173-183

[6] Boschin G, Arnoldi A. Legumes are valuable sources of tocopherols. Food Chemistry. 2011;127(3):1199-1203

[7] Gonçalves A, Goufo P, Barros A, Domínguez-Perles R, Trindade H, Rosa EA, et al. Cowpea (*Vigna unguiculata* L. Walp), a renewed multipurpose crop for a more sustainable

agri-food system: nutritional advantages and constraints. Journal of the Science of Food and Agriculture. 2016;**96**(9):2941-2951

[8]   Rajan K, Ankur T. Nutraceutical and pharmacological properties of Vigna species. Indian Journal of Agricultural Biochemistry. 2017;**30**(1):10-20

[9]   Shimelis AE, Rakshit SK. Anti-nutritional factors and in vitro protein digestibility of improved haricot bean (*Phaseolus vulgaris* L.) varieties grown in Ethiopia. Inter J. Food Science & Nutrition. 2005;**56**:377-387

[10]  Sharma G, Srivastava AK, Prakash D. Phytochemicals of nutraceutical importance: Their role in health and diseases. Pharmacology. 2011;**2**:408-427

[11]  Prakash D, Gupta C. Role of phytoestrogens as nutraceuticals in human health. Pharmacology. 2011;**1**:510-523

[12]  Campos-Vega R, Loarca-Pina G, Dave Oomah B. Minor components of pulses and their potential impact on human health. Foodservice Research International. 2010;**43**:461-482

[13]  Iqbal A, Khalil IA, Ateeq N, Sayyar Khan M. Nutritional quality of important food legumes. Food Chemistry. 2006;**97**(2):331-335

[14]  United Nations (2014): Resolution adopted by the General Assembly on 20 December 2013, 68/231. International Year of Pulses 2016. A / RES / 68/231

[15]  Clifton PM. Legumes and cardiovascular disease. In: Bioactive Foods in Promoting Health. San Diego: Elsevier Inc; 2010. pp. 449-455

[16]  Tharanathan R, Mahadevamma S. Grain legumes–a boon to human nutrition. Trends in Food Science and Technology. 2003;**14**(12):507-518

[17]  Williams PG, Grafenauer SJ, O'Shea JE. Cereal grains, legumes, and weight management: A comprehensive review of the scientific evidence. Nutrition Reviews. 2008;**66**:171-182

[18]  Phillips RD, McWatters KH, Chinnan MS, Hung YC, Beuchat LR, Sefa-Dedeh S, Sakyi-Dawson E, Ngoddy P, Nnanyelugo D, Enwere J, Komey NS, Liu K, Mensa-Wilmot Y, Nnanna IA, Okeke C, Prinyawiwatkul W, Saalia FK. Utilization of cowpeas for human food. Field Crops Research. 2003;**82**:193-213

[19]  Trinidad TP, Mallillin AC, Loyola AS, Sagum RS, Encabo RR. The potential health benefits of legumes as a good source of dietary fibre. British Journal of Nutrition. 2010; **103**(4):569-574

[20]  Shetty AA, Magadum S, Maganvi K. Vegetables as sources of antioxidants. Journal of Food and Nutritional Disorders. 2013;**2**(1):2

[21]  Baptista A, Pinho O, Pinto E, Casal S, Mota C, Ferreira IM. Characterization of protein and fat composition of seeds from common beans (*Phaseolus vulgaris* L.), cowpea (*Vigna unguiculata* L. Walp) and bambara groundnuts (*Vigna subterranea* L. Verdc) from Mozambique. Journal of Food Measurement and Characterization. 2017;**11**(2):442-450

[22] Tarawali SA, Singh BB, Gupta SC. Cowpea as a key factor for a new approach to integrate crop-livestock systems research in dry savannas of West Africa. In: Challenges and opportunities for enhancing sustainable cowpea production. Proceedings of the world cowpea conference III held at the International Institute of Tropical Agriculture (IITA), Ibadan, Nigeria; 2002

[23] Hall A. Phenotyping cowpeas for adaptation to drought. Frontiers in Physiology. 2426;**3**, **2012**:155

[24] Singh BB, Ehlers JD, Sharma B, Freire Filho FR. Recent progress in cowpea breeding. In: Fatokun CA, Tarawali SA, Singh BB, Kormawa PM, Tamò M, editors. Challenges and Opportunities for Enhancing Sustainable Cowpea Production. Ibadan: International Institute of Tropical Agriculture; 2002. pp. 4-8

[25] Antova GA, Stoilova TD, Ivanova MM. Proximate and lipid composition of cowpea (*Vigna unguiculata* L.) cultivated in Bulgaria. Journal of Food Composition and Analysis. 2014;**33**(2):146-152

[26] Langyintuo AS, Lowenberg-DeBoer J, Faye M, Lamber D, Ibro G, et al. Cowpea supply and demand in West Africa. Field Crops Research. 2003;**82**:215-231

[27] Xiong H, Shi A, Mou B, Qin J, Motes D, Lu W, Wu D. Genetic diversity and population structure of cowpea (*Vigna unguiculata* L. Walp). PLoS One. 2016;**11**(8):e0160941

[28] Freire Filho FR, Ribeiro VQ, Barreto PD, Santos AA. Melhoramento genético. In: Freire Filho FR, Lima JAA, Ribeiro VQ, editors. Feijão-caupi: avanços tecnológicos. Brasília: Embrapa Informação Tecnológica; 2005. pp. 27-92

[29] GUERRA JVS, CARVALHO AJD, Medeiros JC, SOUZA AAD, Brito OG. Agronomic performance of erect and semi-erect cowpea genotypes in the north of Minas Gerais, Brazil. Revista Caatinga. 2017;**30**(3):679-686

[30] Draghici R, Draghici I, Diaconu A, Dima M. Variability of genetic resources of cowpea (*Vigna unguiculata*) studied in the sandy soil conditions from Romania. Annals of the University of Craiova-Agriculture, Montanology, Cadastre Series. 2017;**46**(1):147-153

[31] Giller KE. Nitrogen Fixation in Tropical Cropping Systems. Wallingford: CAB International; 2001

[32] Udvardi M, Poole PS. Transport and metabolism in legume-rhizobia symbioses. Annual Review of Plant Biology. 2013;**64**:781-805. DOI: 10.1146/annurev-arplant-050312-120235

[33] Araújo JPP, Watt EE. O caupí no Brasil. IITA/EMBRAPA: Brasília; 1988. p. 722

[34] Rangel OJP, Silva CA. Estoques de carbono e nitrogênio e frações orgânicas de Latossolo submetido a diferentes sistemas de uso e manejo. Revista Brasileira de Ciência do Solo, Viçosa. 2007;**31**(6):1609-1623

[35] Freire Filho FR. Feijão-caupi no Brasil: produção, melhoramento genético, avanços e desafios. 1st ed. Teresina, PI: Embrapa; 2011. p. 84

[36]  Dwivedi SL, van Bueren ETL, Ceccarelli S, Grando S, Upadhyaya HD, Ortiz R. Diversifying food systems in the pursuit of sustainable food production and healthy diets. Trends in Plant Science. 2017;**22**(10):842-856

[37]  Maia FMM, Oliveira JTA, Matos MRT, Moreira RA, Vasconcelos IM. Proximate composition, amino acid content and haemagglutinating and trypsin-inhibiting activities of some Brazilian *Vigna unguiculata* (L) Walp cultivars. Journal of the Science of Food and Agriculture. 2000;**80**(4):453-458

[38]  Davis DW, Delke EA, Oplinger ES, Hanson CV, Putnam DH. Alternative Field Crop Manual, Department of Horticultural Science, Agronomy and Plant Genetics; Center for Alternative Plant and Animals Products. U.S.A: Minnesota Extension Service, University of Minnesota; 2013

[39]  Timko MP, Singh BB. Cowpea, a multifunctional legume. In: Genomics of Tropical Crop Plants. NY: Springer, New York; 2008. pp. 227-258

[40]  Shetty K. Systems Solutions to Global Food Security Challenges to Advance Human Health and Global Environment Based on Diverse Food Ecology Sustainable Investment Produce, 65

[41]  Singh BB. Future Prospects of Cowpea. Cowpea: The Food Legume of the 21st Century, n. cowpeathefoodle; 2014. pp. 145-157

[42]  Tiwari B, Singh N. Pulse Chemistry and Technology. Royal Society of Chemistry; 2012

[43]  Frota KMG, Mendonca S, Saldiva PHN, Cruz RJ, Areas JAG. Cholesterol-lowering properties ofwhole cowpea seed and its protein isolate in hamsters. Journal of Food Science. 2008;**73**:H235-H240

[44]  Kapravelou G, Martinez R, Andrade AM, Chaves CL, Lopez-Jurado M, Aranda P, et al. Improvement of the antioxidant and hypolipidaemic effects of cowpea flours (*Vigna unguiculata*) by fermentation: results of in vitro and in vivo experiments. Journal of the Science of Food and Agriculture. 2015;**95**:1207-1216

[45]  Xiong SL, Yao XL, Li AL. Antioxidant properties of peptide from cowpea seed. International Journal of Food Properties. 2013;**16**:1245-1256

[46]  Xu BJ, Chang SKC. Comparative study on antiproliferation properties and cellular antioxidant activities of commonly consumed food legumes against nine human cancer cell lines. Food Chemistry. 2012;**134**:1287-1296

[47]  Bello SK, Yusuf AA, Cargele M. Performance of cowpea as influenced by native strain of rhizobia, lime and phosphorus in Samaru, Nigeria. Symbiosis. 2017;**35**:1-10

[48]  Bialostosky K, Wright JD, Kennedy-Stephenson J, McDowell M, Johnson CL. Dietary intake of macronutrients, micronutrients, and other dietary constituents: United States, 1988-94: data from the National Health Examination Survey, the National Health and Nutrition Examination Surveys, and the Hispanic Health and Nutrition Examination Survey; 2002

[49] Giami SY. Compositional and nutritional properties of selected newly developed lines of Cowpea (*Vigna unguiculata* L. Walp). Journal of Food Composition and Analysis. 2005;**18**:665-673

[50] Carvalho AFU, de Sousa NM, Farias DF, da Rocha-Bezerra LCB, da Silva RMP, Viana MP, de Morais SM. Nutritional ranking of 30 Brazilian genotypes of cowpeas including determination of antioxidant capacity and vitamins. Journal of Food Composition and Analysis. 2012;**26**(1-2):81-88

[51] Nielsen SS, Brandt WE, Singh BB. Genetic variability for nutritional composition and cooking time of improved cowpea lines. Crop Science. 1993;**33**:469-472

[52] Santos CAF, Boiteux LS. Breeding biofortified cowpea lines for semi-arid tropical areas by combining higher seed protein and mineral levels. Genetics and Molecular Research. 2013;**12**(4):6782-6789

[53] Duranti M. Grain legume proteins and nutraceutical properties. Fitoterapia. 2006;**77**:67-82

[54] Boye J, Wijesinha-Bettoni R, Burlingame B. Protein quality evaluation twenty years after the introduction of the protein digestibility corrected amino acid score method. The British Journal of Nutrition. 2012;**108**(S2):S183-S211

[55] Pandurangan S et al. Differential response to sulfur nutrition of two common bean genotypes differing in storage protein composition. Frontiers in Plant Science. 2015;**6**:92

[56] Frota KDMG, Lopes LAR, Silva ICV, Arêas JAG. Nutritional quality of the protein of *Vigna unguiculata* L. Walp and its protein isolate. Revista Ciência Agronômica. 2017; **48**(5SPE):792-798

[57] Vasconcelos IM et al. Protein fractions, amino acid composition and antinutritional constituents of high-yielding cowpea cultivars. Journal of Food Composition and Analysis. 2010;**23**(1):54-60

[58] Food and Agriculture Organization; World Health Organization. Report of a joint FAO/WHO expert consultation. Protein Quality Evaluation. Pomegranate; 1991. p. 66 (FAO Food and Nutrition Paper, 51)

[59] Anjos F et al. Chemical composition, amino acid digestibility, and true metabolizable energy of cowpeas as affected by roasting and extrusion processing treatments using the cecectomized rooster assay. The Journal of Applied Poultry Research. 2016;**25**(1):85-94

[60] Elhardallou S et al. Amino acid composition of cowpea (Vigna ungiculata L. Walp) flour and its protein isolates. Food and Nutrition Sciences. 2015;**6**(9):790-797

[61] Kadam SS, Salunkhe DK. Nutritional composition, processing, and utilization of horse gram and moth bean. Critical Reviews in Food Science and Nutrition. 1985;**22**:1-26

[62] Sarwar G, Xiao C, Cockell KA. Impact of antinutritional factors in food proteins on the digestibility of protein and the bioavailability of amino acids and on protein quality. The British Journal of Nutrition. 2012;**108**(S2):S315-S332

[63]  Elhardallou SB, Khalid II, Gobouri AA, Abdel-Hafez SH. Amino acid composition of cowpea (Vigna ungiculata L. Walp) flour and its protein isolates. Food and Nutrition Sciences. 2015;**6**(9):790

[64]  Hoover R, Hughes T, Chung HJ, Liu Q. Composition, molecular structure, properties and modification of pulse starches: A review. Food Research International. 2010;**43**:399-413

[65]  Ashogbon AO, Akintayo ET. Isolation and characterization of starches from two cowpea (*Vigna unguiculata*) cultivars. International Food Research Journal. 2013;**20**:3093-3100

[66]  Ratnaningsih N, Suparmo EH, Marsono Y. Composition, microstructure and physico-chemical properties of starches from Indonesian cowpea (*Vigna unguiculata*) varieties. International Food Research Journal. 2016;**23**:2041-2049

[67]  Yamada Y, Hosoya S, Nishimura S, Tanaka T, Kajimoto Y, Nishimura A, Kajimoto O. Effect of bread containing resistant starch on postprandial blood glucose levels in humans. Bioscience, Biotechnology, and Biochemistry. 2005;**69**:559-566

[68]  Park OJ, Kang NE, Chang MJ, Kim WK. Resistant starch supplementation influences blood lipid concentrations and glucose control in overweight subjects. Journal of Nutritional Science and Vitaminology. 2004;**50**:93-99

[69]  Thorne MJ, Thompson LU, Jenkins DJ. Factors affecting starch digestibility and the glycemic response with special reference to legumes. The American Journal of Clinical Nutrition. 1983;**38**:481-488

[70]  Foster-Powell K, Holt SH, Brand-Miller JC. International table of glycemic index and glycemic load values. The American Journal of Clinical Nutrition. 2002;**76**:55-56

[71]  Du SK, Jiang H, Ai Y, Jane JL. Physicochemical properties and digestibility of common bean (*Phaseolus vulgaris* L.) starches. Carbohydrate Polymers. 2014;**108**:200-205

[72]  Martínez-Villaluenga C, Frías J, Vidal-Valverde C. Alpha-Galactosides: antinutritional factor or functional ingredients? Critical Reviews in Food Science and Nutrition. 2008;**48**:301-316

[73]  Embrapa Meio-Norte. Cultivo de feijão caupi. Jul/2003. Disponível em:. Acesso em: 8 mar. 2007

[74]  Frota KDMG, Soares RAM, Arêas JAG. Composição química do feijão caupi (*Vigna unguiculata* L. Walp), cultivar BRS-Milênio. Ciência e Tecnologia de Alimentos. 2008; **28**(2):470-476

[75]  Antova GA, Stoilova TD, Ivanova MM. Proximate and lipid composition of cowpea (*Vigna unguiculata* L.) cultivated in Bulgaria. Journal of Food Composition and Analysis. 2014;**33**:146-152

[76]  Zia-Ul-Haq M, Ahmad S, Chiavaro E, Mehjabeen AS, Sagheer A. Studies of oil from cowpea (*Vigna unguiculata* (L.) Walp) cultivars commonly grown in Pakistan. Pakistan Journal of Botany. 2010;**42**:1333-1341

[77]  DellaPenna D. Nutritional genomics: manipulating plant micronutrients to improve human health. Science. 1999;**285**(5426):375-379

[78]  Asare AT, Agbemafle R, Adukpo GE, Diabor E, Adamtey KA. Assessment of functional properties and nutritional composition of some cowpea (*Vigna unguiculata* L.) genotypes in Ghana. Journal of Agricultural and Biological Science. 2013;**8**(6):465-469

[79]  HARDISSON A et al. Mineral composition of the banana (*Musa acuminata*) from the island of Tenerife. Food Chemistry. 2001;**73**:153-161

[80]  Santos CAF, Boiteux LS. Breeding biofortified cowpea lines for semi-arid tropical areas by combining higher seed protein and mineral levels. Genetics and Molecular Research. 2013;**12**(4):6782-6789

[81]  Welch RM, Graham RD. Breeding for micronutrients in staple food crops from a human nutrition perspective. Journal of Experimental Botany. 2004;**55**:353-364

[82]  Kokate CK, Purohit AP, Gokhale SB. Nutraceutical and Cosmaceutical. Pharmacognosy. 21st ed. Pune, India: Nirali Prakashan; 2002. pp. 542-549

[83]  Kalia AN. Textbook of Industrial Pharmacognosy. New Delhi: CBS Publisher and Distributor; 2005. pp. 204-208

[84]  Roberfroid M. Health benefits of non-digestible oligosaccharides. In: Kritchevsky D, Bonfield C, editors. Dietary Fiber in Health and Disease (Advances in Experimental Biology). New York: Plenum Press; 1997. p. 427

[85]  Anderson JW, Chen WJL. Cholesterol-lowering properties of oat products. In: Webster FH, editor. Oats Chemistry and Technology. St Paul, MN: American Association of Cereal Chemists; 1986. pp. 309-333

[86]  Sreerama YN, Sashikala VB, Pratape VM, Singh V. Nutrients and antinutrients in cowpea and horse gram flours in comparison to chickpea flour: Evaluation of their flour functionality. Food Chemistry. 2012;**131**(2):462-468

[87]  Rolfe RD. The role of probiotic cultures in the control of gastrointestinal health. Journal of Nutrition. 2000;**130**:396S-402S

[88]  Fuller R. Probiotics in man and animals. Journal of Applied Bacteriology. 1989;**66**:365-378

[89]  Lei V, Jakobsen M. Microbiological characterization and probiotic potential of koko and koko sour water, an African spontaneously fermented millet porridge and drink. Journal of Applied Microbiology. 2004;**96**:384-397

[90]  Vasiljevic T, Shah NP. Probiotics. From Metchnikoff to bioactives. International Dairy Journal. 2008;**18**:714-728

[91]  Schaffner DW, Beuchat LR. Fermentation of aqueous plant seed extracts by lactic acid bacteria. Applied and Environmental Microbiology. 1986;**51**:1072-1076

[92]  Gibson GR, Roberfroid MB. Dietary modulation of the human colonic microbiota: introducing the concept of prebiotics. The Journal of Nutrition. 1995;**125**:1401-1412

[93]  Swennen K, Courtin CM, Delcour JA. Non-digestible oligosaccharides with prebiotic properties. Critical Reviews in Food Science and Nutrition. 2006;**46**:459-471

[94]  Salminen S, Bouly C, Boutron-Ruault MC, Cummings JH, Franck A, Gibson GR, Isolauri E, Moreaou MC. Functional food science and gastrointestinal physiology and function. The British Journal of Nutrition. 1998;**80**:S147-S171

[95]  Roberfroid M. Functional food concept and its application to prebiotics. Digestive and Liver Disease. 2002;**34**:S105-S110

[96]  Sreerama YN, Sashikala VB, Pratape VM, Singh V. Nutrients and antinutrients in cowpea and horse gram flours in comparison to chickpea flour: Evaluation of their flour functionality. Food Chemistry. 2012;**131**(2):462-468

[97]  Martínez-Villaluenga C, Frías J, Vidal-Valverde C. Alpha-galactosides: antinutritional factors or functional ingredients? Critical Reviews in Food Science and Nutrition. 2008;**48**(4):301-316

[98]  Swennen K, Courtin CM, Delcour JA. Non-digestibleoligosaccharides with prebiotic properties. Critical Reviews in Food Science and Nutrition. 2006;**46**:459-471

[99]  Ruiz-Aceituno L, Rodríguez-Sánchez S, Ruiz-Matute AI, Ramos L, Soria AC, Sanz ML. Optimisation of a biotechnological procedure for selective fractionation of bioactive inositols in edible legume extracts. Journal of Science and Food Agriculture. 2013;**93**: 2797-2803

[100]  Escott-Stump E, Mahan LK. Krause's Food, Nutrition and Diet Therapy. 10th ed. Philadelphia: WB Saunders Company; 2000. pp. 553-559

[101]  Onwuliri VA, Obu JA. Lipids and other constituents of *Vigna unguiculata* and *Phaseolus vulgaris* grown in northern Nigeria. Food Chemistry. 2002;**78**:1-7

[102]  Hashim N, Pongjata J. Vitamin A activity of rice-based weaning foods enriched with germinated cowpea flour, banana, pumpkin and milk powder. Malaysian Journal of Nutrition. 2000;**6**:65-73

[103]  Kalogeropoulos N, Chiou A, Ioannou M, Karathanos VT, Hassapidou M, Andrikopoulos NK. Nutritional evaluation and bioactive microconstituents (phytosterols, tocopherols, polyphenols, triterpenic acids) in cooked dry legumes usually consumed in the Mediterranean countries. Food Chemistry. 2010;**121**:682-690

[104]  Tsuda T, Makino Y, Kato H, Osawa T, Kawakishi S. Screening for antioxidative activity of edible pulses. Bioscience, Biotechnology, and Biochemistry. 1993;**57**:1606-1608

[105]  Apea-Bah FB, Serem JC, Bester MJ, Duodu KG. Phenolic composition and antioxidant properties of koose, a deep-fat fried cowpea cake. Food Chemistry. 2017;**237**:247-256

[106] Manach C, Scalbert H, Morand C, Remesy C, Jimenez L. Polyphenols in foods and bioavailability. The American Journal. Clinical Nutrition. 2004;**79**(5):727-747

[107] Wu X, Beecher GR, Holden JM, Haytowitz DB, Gebhardt SE, Prior RL. Concentrations of anthocyanins in common foods in the United States and estimation of normal consumption. Journal of Agricultural and Food Chemistry. 2006;**54**(11):4069-4075

[108] Soobrattee MA, Neergheen VS, Luximon-ramma A, Aruomab OI, Bahorun T. Phenolics as potential antioxidant therapeutic agents: Mechanism and actions. Mutation Research. 2005;**579**(1-2):200-213

[109] McDougall GJ, Stewart D. The inhibitory effects of berry polyphenols on digestive enzymes. BioFactors. 2005;**23**(4):189-195

# Unconventional Legumes

# Chapter 4

# Prosopis cineraria as an Unconventional Legumes, Nutrition and Health Benefits

Hanan Sobhy Amin Afifi and Ihsan Abu Al-rub

Additional information is available at the end of the chapter

http://dx.doi.org/10.5772/intechopen.79291

## Abstract

*Prosopis cineraria* (L.) Druce is considered as one of the highly valued plants in the native system of medicine for many arid and dry areas in the world. Ancient literature for Arabian Gulf and Indian desert illustrated the important of the plant in treated various ailments like asthma, dysentery, leucoderma, leprosy, dyspepsia, earache, etc. The present chapter review the using of *P. cineraria* as unconventional legumes that not well known as a rich and sustainable source of protein for many people in the world. It emphasis on its broad food and nonfood applications, nutritional values and health benefits. As well as looking at the phytochemical constituent's content that has been identified in the various parts of the plant as alkaloid, steroids, alcohol and alkane. The present paper describes the morphological trait of *P. cineraria* and identifies the environmental conditions required for its natural distribution. Historically, this plant has drag attention for its various uses therefore, it has been considered as the National Tree of the United Arab Emirates in the Arabian Gulf.

**Keywords:** *Prosopis cineraria*, Leguminosae, nutritional value, pharmacological properties, usage, phytochemicals

## 1. Introduction

The continuous world population growth, inadequate protein sources, exorbitant cost of animal protein are considered the main reasons for malnutrition and undernourishment among people living in many developing countries around the world. To meet the increasing demand of protein, alternative strategies and unconventional sources of protein for human and animal nutrition have been considered recently.

Trees of *Prosopis* genus, which belongs to the Leguminosae family, are one of the most important source of proteins in arid and semi-arid regions. Its capability to stand heat and tolerate drought, salt, and alkalinity make *Prosopis* cultivated and distributed in many areas around the world especially India, America, GCC, and MENA [1]. According to the recent studies, the species *Prosopis cineraria* has significant contribution in the farm economy and rural area development. Undoubtedly, it shares with other *Prosopis* species numerous characteristics, uses and effects, i.e., chemical composition, types of phytochemical components, and health effects. *Prosopis cineraria* has been valued by different communities and cultures for the versatility of all its parts and named as "the Wonder Tree" or "King of Desert" [2] or "the Golden Tree of Indian deserts" [3]. The tree parts including leaves, pods, seeds and barks has been used in many ways as food, i.e., flour, drink, vegetable, and gum. Leaves and pods are used for ruminant and animal feed. *Prosopis cineraria* extensively used in traditional medicine to cure many diseases such as ailments like leprosy, dysentery, asthma, leucoderma, dyspepsia and earache [4–6]. Barks are used for non-nutritional purposes, i.e., wood, tanning, fuel, firewood and charcoal. The *Prosopis cineraria* has many chemical constituents as alkaloid, steroids, alcohol and alkane.

Despite its fabulous importance in local culture, there is minimal aware by the developed communities about *P. cineraria* as unconventional legumes. Therefore, authors present a comprehensive chapter about this important tree from all aspects including traditional uses, biological and phytochemical investigation.

## 2. Botany

The genus *Prosopis* L. belongs to Leguminosae family, subfamily Mimosoideae and accommodates 44 species of which 40 are native to North and South Americas, three originate in Asia, and one comes from Africa [7–9]. Trees of *Prosopis* L. are widespread in Western Asia, Africa and arid and semi-arid regions in the Americas and Australia.

The species *P. cineraria* is native to dry and arid regions of Arabia and India [10]. Its main population is center on the Thar Desert of India and Pakistan, with less dense populations occur in the Arabian Peninsula, Iran, and Afghanistan [11]. It is considered the national tree of the United Arab Emirates [12]. *P. cineraria* is known as Ghaf in Arabic, Khejri in Indian, and Jand in Pakistan.

*P. cineraria* is an evergreen, thorny tree, 10–25 m in high. The stem is commonly straight, unbranched for several meters with a gray roughish, exfoliated bark (**Figure 1**). The branches are slender, drooped giving the canopy a rounded appearance with short triangular spines (3–6 mm long) between leaves nodes. At the time of no grazing the lower branches can reach to the ground. Leaves are gray-green, alternate usually divided into two pinnae, each pinna has 7–14 pairs of oblong, oblique, apex leaflets. The mid-rib nearer the upper edge, is sessile.

Flowers are small, yellow or creamy white, nearly sessile in slender pedunculated axillary spikes 5–13 cm long. Pods are yellow to reddish brown with cylindrical shape and slightly curved; 10–20 cm long and 0.5–0.8 cm thick. Seeds 10–25, oblong or rhomboidal,

**Figure 1.** The tree of *Prosopis cineraria*, flower, leave and pods.

brown, smooth, with a moderately hard taste [13, 14]. The tap root of *P. cineraria* penetrates vertically up to 20 m but can reach water at an extraordinary depth of 53 m or more [15]. Flowering and fruiting period is varied between locations and weather condition and generally from February to May after the new flush of leaves. The pods are mature almost after 2 months.

## 3. Environmental conditions

*P. cineraria* is a xerophytic plant that is well adapted to dry and arid environment. Under the conditions of drought, the tree produces more flowers and fruits [16]. In areas of its natural distribution, the annual rainfall ranges between 100 up to 500 mm annually, whereas the optimum density is confined to areas receiving 350–400 mm [17]. The climate is characterized by extremes summer temperature varies from about 40–48°C [18]. It can tolerate frost and withstand low temperature less than 10°C in the winter season.

The tree grows on a variety of soils. It is seen at its best on alluvial soils consisting of various mixtures of sand and clay [19]. In arid areas, the growth is better in dune lows than in sandy plains. Good drainage is very essential. *P. cineraria* can grow under highly saline and alkaline soils. However, it relatively salt tolerant at seed germination whereas seedling emergence was found to be reduced to 50% in soil with a salinity of 7.6 dS m⁻¹ and a further increase in salt concentration was detrimental to seed germination [20].

## 4. Socio economic and ecological importance

*P. cineraria* is a multipurpose tree that holds an important role in the rural economy in many arid regions, particularly in the Arabian Gulf and the northwest arid region of Indian sub-continent. Historically, the Bedouin and Indian uses all its part in their traditional life-style [21–23]. It is used as a folk remedy for various diseases and conditions [24].

The unripe pods are used for making curry and pickle. The green pods are consumed as vegetables. The flour of mature pods is used for cookies preparation and other local dishes. The leaves and dry pods are annually harvested for cattle and sheep feed, where an adult tree produce 2–5 kg/year dry pods. A resin occurring naturally on the tree, known as mesquite gum, is also occasionally eaten by people [25].

*P. cineraria* as a leguminous tree has importance in improving soil fertility through fixing atmospheric nitrogen. Litter fall production for *P. cineraria* and decomposition rate are con-sidered the highest comparing with other arid trees, and that build up soil organic matter contents under its canopy, increase soluble calcium and available phosphorus and decrease soil pH [26, 27]. Therefore, farmers tend to grow field crops under its canopy to boost the growth and productivity of their crops.

The rounded shape crown provides the shade and shelter for animals and wildlife during hot season. It is widely used for sand dune stabilization program because of it is deep mass root system which enable plant not to compete with others for moisture and nutrients [28]. It provides good quality resources of wood for basic construction and fuel for people in the desert regions.

*P. cineraria* is one of major bee foraging plant in the Arabian Gulf [29], it supports honey bees with long and abundant flowering and honey produced is of a good quality.

## 5. Nutritional value

Numerous people around the world, especially in Africa and Asia, are suffering from protein deficiency due to lack of protein-rich food. *P. cineraria* have 16.5–18.25% protein content com-pared with 25.47% in *Acacia nilotica* and 38.89% in *Acacia senegal* [30]. On other hand, legumes contain 18–35% protein [31], and cereals contain 10–15% protein [32]. Therefore, *Prosopis* seeds are considered a potential and cheap source of protein for industrial use, especially in developing Afro–Asian countries and can be an alternate protein source for solving the protein-energy-malnutrition problem. The protein content, *P. cineraria* contains reasonable amount of ash (5.34%), and fiber (20.93%) [33–35]. Chemical composition of pods is varied between individual trees that it influenced by a wide range of environmental factors. The *P. cineraria* pods have low moisture content (8.55%) that may be advantageous in increasing of the pods shelf-life, 18% protein, 1.89% oil, 5.34% ash and 20.93% fiber [34]. The *P. cineraria* seed contains 10.6% oil, 28.6% of the oil are saturated fatty esters, 68.3% are unsaturated fatty esters, and 3.1% are methyl hydroxy fatty ester. Moreover, the seed oil is rich in oleic acid (31.3%) along with linoleic acid (32.1%). Oil and seeds of *P. cineraria* show an absence of keto, cyclopropenoid, and epoxy fatty acids or any evidence for the presence of trans-unsaturation or the presence of conjugation. In addition, the tree leaves have a good source of macro miner-als as calcium (2.43%), phosphorus (0.16%) and potassium (0.41%). So, it can be used as good food during the mineral deficient periods [36].

# 6. Usage

Besides the ecological value of *P. cineraria* tree, there are significant utilizations centered on its use for human food, animal feeds, medical purposes and many other applications. The multipurpose and added value usages of *P. cineraria* tree; barks, pods, and leaves; will be discussed with regards to its health benefits and nutraceutical effects as follow:

## 6.1. Human nutrition/food application

*P. cineraria* tree are extensively used as human food in many area especially arid land region and semi-desert as Arizona, India, California, South America and northwestern Mexico. There are diverse uses of the *P. cineraria* tree parts; dried and undried pods, green and dry leaves, and seeds; in human food. It is interesting to note that studies did not refer to the presence of cyanogenic or toxic compounds in *Prosopis* parts as seeds or pods till now [37–39]. The *P. cineraria* food applications include:

### 6.1.1. Vegetables

Leguminous *Prosopis* trees play a great role in feeding human in dry area to prevent protein and mineral deficiency especially during famine period. In these area, people used to eat unripe green pods of *P. cineraria* that selling in their market as vegetables and children eat its ripe fruits [2, 33, 40, 41]. In addition, green and unripe pods are also used in the preparation of pickles and curries [3].

### 6.1.2. Flour

The *Prosopis* pods consist of three parts, mesocarp (56% of the pod) that grind to produce flour, endocarp (35%) that discard as waste alongside seeds (9%). People used the flour to make bread, cake, chapatti by mixing with wheat flour and sweets [40]. The *Prosopis* flour contains a high level of proteins (62%), dietary fiber (25%) and low content of total carbohydrate and fat in addition to dominant amounts of free polyphenol and carotenoids compounds as shown in **Table 1** [42]. *Prosopis* flour is gluten-free, and a premium source of calcium, potassium, magnesium, zinc, and iron, in addition to amino acids such as lysine that is low in other cereals [11, 43]. *Prosopis* flour has a unique combination taste that has been variously described as; sweet or slightly nutty, with a sweet chocolate or coffee flavor, with a pleasant hint of caramel or molasses, with a hint of cinnamon as it contains many volatile components, i.e., $\gamma$-nonalactone, 5,6-dihydro-6-propyl-2H-pyran-2-one, 2,6-dimethylpyrazine, and methyl salicylate [44]. Therefore, only 10 to 25% flour is generally used in combination with other flours because above than 25%, the taste becomes too strong for most palate. While, the desirable degree of browning for different bakery products was obtained using different adding concentration, i.e., biscuits (5%), breads (10%), pancakes (15%) and chapatti (50%).

The dried pods are used to make flour after collecting pods directly from the tree or from pods that have recently fallen to the ground. Sometimes they store the dried pods to provide food year round. The flour particle size is varied depending on the grinding processing,

| Compounds | *Prosopis* flour | Plain white wheat flour |
|---|---|---|
| Energy (kcal/100 g) | 361 | 338 |
| Carbohydrate (g/100 g) | 69.2 | 72.2 |
| Total sugars (g/100 g) | 13.0 | 1.5 |
| Fiber (g/100 g) | 47.8 | 3.2 |
| Protein (g/100 g) | 16.2 | 9.4 |
| Fat content (g/100 g) | 2.12 | 1.3 |
| Saturated fatty acids (g/100 g) | 0.6 | 0.2 |

**Table 1.** Nutritional values of *Prosopis* flour compared with plain white wheat flour.

e.g., pounded using pestle and mortar produces coarse powder, while using stone grinding produces a fine powder.

The *Prosopis* flour assist the diabetic patient through helping maintain a healthy insulin system in those people not affected by blood sugar troubles because of two reasons: firstly, the *Prosopis* flour requires a longer time to be digested compared with other grains, i.e., it needs 4 to 6 hours compared to 1 to 2 hours needs for wheat flour to be digest. This help to sustains constant blood sugar over time and prevents hunger. Secondly, the pods contain fructose, which the body can process without insulin [45].

### 6.1.3. Syrup and drinks

In many places, the *Prosopis* species are used to make fermented, non-fermented beverages, and syrup [46–50]. Nutritious syrup is produced by boiling the clean green pods in water after breaking them into small pieces. Beans should be simmered for 2 hours with continuous adding a small amount of water to avoid burning. Followed by mashing the pods to release more of the sweet pulp with simmering for further few minutes. The juice then sieved through strain and kept in clean containers to be used directly as a drink. Or more sugar can be added to the juice and boil to produce unique flavor syrup [51].

### 6.1.4. Gum

In addition to the previous uses, amber colored gum is produced from the *P. cineraria* tree. This gum has similar properties to the gum produced from acacia tree [40]. Its exudate gum is liquid, water soluble and slowly hardening. Moreover, this genus is not the only source of gum. A galactomannan types interesting gum that called vinal gum is produced from *P. ruscifolia* [52].

### 6.2. Animal nutrition

*P. cineraria* is an important feed species under traditional livestock production systems in the arid regions. Leaves and pods are highly palatable, nutritious and eaten readily by camels, cattle, sheep and goats.

The leaves contained 12.1% crude protein, 20.1% crude fiber, 3.2% ether extract and 12.2% ash [53]. The ripened pods contained 91% dry matter, 13.5% crude protein, 14.3% crude fiber, 1.3% ether extract and 5.2% ash [54]. Feeding *P. cineraria* to sheep did not cause overt health problems such as diarrhea or impaction. Though, it is not advisable to use leaves as a sole feed for animal as it contain 8–10% tannins [55]. Increasing *Prosopis* tannin in the diet reduce animal intake, digestibility of nutrients and body weight gain in sheep [54, 56] and goats [57]. In general goat showed superior efficiency in utilizing *P. cineraria* leaves than that in sheep [58]. However, feeding *Prosopis* tannin at 23 and 45 g/kg dry matter in the ration of lambs and kids can achieve maximum microbial protein synthesis under intensive feeding system. Beyond this level, *Prosopis* tannins will have anti-nutritional effects [59].

## 6.3. Health benefits

Despite the economic importance of *Prosopis* spp. as food, plants have been used in traditional medicine to treat various human ailments since ancient history. *Prosopis* spp. is one of these plants that possess many medicinal properties and used to cure many diseases. Studies showed that leaves and seeds were largely used to treat many diseases such as diarrhea, inflammation, measles, diabetes and prostate disorders [4, 5].

The pods of *P. cineraria* contain alkaloids (good anesthetic and spasmolytic activity), Saponin (boost immunity system of the body, lowering the cholesterol level in the body and reducing the risk of intestinal cancer), and tannins (produce anthelmintic activity). In addition to the mineral content as zinc (relevant to the nutritional aspect as zinc supplementation in diabetes mellitus have antioxidant effect), magnesium (important for proper functioning of every organ like heart, muscle, and kidney), iron (used in anemia, tuberculosis and growth disorder), calcium and phosphorous (useful for the bone, teeth, and ligament related disorder) [17, 60].

Moreover, studies show that the alkaloid mixture of *P. cineraria* in a dose of 1 mg/kg decreased the blood pressure and immediate mortality of dogs. In contrast, extensive damage to the liver, spleen, kidney, lung, and heart was observed on histological examination of mice given the same alkaloid mixture [61].

### 6.3.1. Antimicrobial activity

Studies show that the methanolic extract of *Prosopis* pods has antimicrobial activity against *Candida albicans* [62]. And the aqueous and methanolic extracts of stem bark have moderate antibacterial activity at a dose of 250 µg/ml. In addition to the previous effects, the methanolic extract shows significant action on all pathogens. This antibacterial activity of *Prosopis* spp. is due to the presence of flavonoids and tannins [63].

### 6.3.2. Antihyperglycemic (antidiabetic) and antioxidant activities

Many researchers illustrated that the bark extract of the *P. cineraria* have abundant activity in lowering blood sugar level by 27.3%, in addition, to significant decrease in body weight (29.6%) in diabetic rats when a dose of 300 mg/Kg mice body weight are given orally in daily base for 45 days [6, 64] explained the effect of the *Prosopis* extracts is due to activate the surviving of the β cells of the islets of langerhans and producing an insulinogenic effect.

### 6.3.3. Antihypercholesterolemic activity

The 70% hydroalcoholic bark extract dose of 500 mg/Kg BW of albino male New Zealand white rabbits reduced significantly the serum total cholesterol by 88%, LDL-C by 95%, triglyceride by 59%, VLDL-C by 60% and ischemic indices compared to hypercholesterolemic control [64–66].

### 6.3.4. Antitumor activities

A study on *P. cineraria* illustrated that a dose of 200 and 400 mg/Kg BW of hydroalcoholic extract of leaves and bark have a significant antitumor activity against Ehrlich ascites carcinoma tumor model. In addition, the methanolic extract of the *P. cineraria* leaves shows significant radical scavenging activity. This effect is due to the inhibition of cell proliferation even through inducing the cell death and/or extending the time for cell proliferation [67].

### 6.3.5. Antidepressant effect

Studies show that aqueous extract of the *P. cineraria* leaves have a significant antidepressant effect on mice and a similar effect of the antidepressant drugs. This is due to the presence of some phytochemicals as saponins, flavonoids, glycosides, alkaloids, and phenolic compounds in these extracts [5].

### 6.3.6. Toxicity studies

Toxicity effect of 50% Hydroalcoholic extracts of *Prosopis* (at dose ranged between 50 and 2000 mg/Kg BW) through oral route of rats did not show any significant effects in breathing, behavior, sensory nervous system responses, cutaneous effects or had any mortality recorded within 24 h after treatments [6]. Further studies are required to determine the toxicity effects of the *Prosopis* extracts that might show adverse effects when consumed because it contains piperidine alkaloids [68].

## 6.4. Other uses

These are not the only uses of the *P. cineraria* tree. The good bark considered a good source of woods that can be used to make tool handles, boat frames, posts, and houses. While the poor or bad quality bark can be used as timber [40]. In India; especially in the Punjab region; the purplish brown bark used as fuel, firewood and used to produce high-quality charcoal. Leaf galls of *P. cineraria* tree can be used also for tanning. While, leaves can be used as a source of compost on the agricultural field and flowers are considered a good source for honey bee forage. The produced honey is light yellow with pleasant taste and slight aroma and generally of good quality [69].

# 7. Phytochemicals

There are few studies on the chemistry and bioactive compounds of *Prosopis* species have been published recently. Studies referred to the secondary metabolites compounds in plants that are considered bioactive compounds and has diverse antinutritional and nutraceutical features. Therefore, it can be potential as a source of bioactive products and used in functional products. Refs. [61, 84–86] mentioned that *Prosopis* spp. tree generally contains various phytochemical compounds as tannins, 5-hydroxytryptamine, isorhamnetin-3-diglucoside, L-arabinose, quercetin, apigenin, and tryptamine. Studies conducted on phytochemical compounds of *P. cineraria* showed that each part of the plant contains different types of these compounds (**Tables 2** and **3**).

| Plant part | Chemical constituent present | Medicinal effect |
|---|---|---|
| Flowers | Patuletin glycoside patulitrin, luteolin and rutin sitosterol, and spicigerine.<br><br>Flavone derivatives Prosogerin A and Prosogerin B | -Flowers are known as an anti-diabetic agent.<br><br>-Flowers can be mixed with sugar when administered orally prevent miscarriage.<br><br>-It contains Patulitrin3, 5, 6, 3, 4-pentamethoxy-7-hydroxy flavone which has significant activity against Lewis lung carcinoma in vivo.<br><br>**References:** [70, 71] |
| Leaves | -Alkaloid: spicigerine<br><br>-Steroids: campesterol, cholesterol, sitosterol, stigmasterol, actacosanol<br><br>-Alcohol: octacosanal, triacontane-1-ol, Tricosan-1-ol,<br><br>-Alkane: hentriacontane, Diisopropyl-10,11-dihydroxyicosane-1,20-dioate<br><br>**References:** [6, 72–76] | -Leaf paste of *P. cineraria* is applied on boils and blisters, including mouth ulcers in livestock and leaf infusion on open sores on the skin<br><br>-Smoke of the leaves is considered good for eye troubles and infections.<br><br>**References:** [72, 77–80] |
| Seeds | Prosogerin C, Prosogerin D, Prosogerin E, gallic acid, patuletin, patulitrin, luteolin, and rutin | |
| Pods | 3-benzyl-2-hydroxy-urs-12-en-28-oic acid, maslinic acid-3 glucoside, linoleic acid, prosophylline, 5,5'-oxybis-1,3-benzenediol, 3,4,5-trihydroxycinnamic acid 2-hydroxyethyl ester and 5,3',4'trihydroxyflavanone 7-glycoside | -Dry pods help in preventing protein calorie malnutrition and iron calcium deficiency in blood.<br><br>**References:** [3, 64] |
| Barks | Hexacosan-25-on-l-ol, a new keto alcohol along with ombuin and a triterpenoid glycoside.vitamin K1, n-octacosyl acetate, the long-chain aliphatic acid.<br><br>Presence of glucose, rhamnose, sucrose and starch | -Bark used in the treatment of asthma, bronchitis, dysentery, leucoderma, leprosy, muscle tremors and piles.<br><br>-Different extracts of stem bark possessed a weak antibacterial activity.<br><br>**References:** [81–83] |

**Table 2.** Phytochemical constituents of the ***Prosopis cineraria***.

| Phytochemicals | Flower | | Leaf | | Pod | | Seed | | Stem | |
|---|---|---|---|---|---|---|---|---|---|---|
| Plant parts | Aqueous | Ethanol | Aqueous | Ethanol | Aqueous | Ethanol | Aqueous | Ethanol | Aqueous | Ethanol |
| Carbohydrates | + | + | – | – | +++ | +++ | + | + | – | – |
| Proteins | – | – | + | + | ++ | ++ | + | + | – | – |
| Tannin | – | – | + | + | + | + | + | + | – | – |
| Flavonoids | ++ | +++ | + | + | + | ++ | ++ | ++ | + | – |
| Cardia glycoside | – | – | – | – | + | + | – | – | – | – |
| Alkaloids | ++ | ++ | ++ | ++ | ++ | ++ | ++ | + | – | – |
| Terpenes | – | – | + | + | + | + | + | ++ | + | + |
| Steroids | + | + | +++ | +++ | – | – | + | ++ | – | – |

+; low concentration, ++; moderate concentration, +++; high concentration, –; absent.

**Table 3.** Concentration of phytochemicals of different parts of *Prosopis cineraria* among different solvents (water and ethanol extracts) [87].

# 8. Conclusions

*P. cineraria* is a naturalized constituent of many natural and cultivated ecosystems in the world. Its value, however, lies not only in its ability to thrive under adverse conditions, but also it provide wide range of useful product. In this unifying review, it was shown the morphological trait, ecological and economical importance in addition to the nutritional value and health benefits.

The authors tried to drag the attention toward this significant tree as alternative type for the traditional legumes and possibility to use it as a source of protein in free-gluten products and functional foods which can be added value in food product development.

Future efforts are required to be focus on integrated management of *P. cineraria* in their natural ecosystem and implement environmental conservation strategies for achieving sustainable uses and maintain its benefits to livelihood and coming generation.

# Acknowledgements

Authors would like to thank Abu Dhabi Food Control Authority, UAE.

# Conflict of interest

The authors declare that there are no conflicts of interest regarding the publication of this chapter.

# Author details

Hanan Sobhy Amin Afifi* and Ihsan Abu Al-rub

*Address all correspondence to: hanan.afifi@adfca.ae

Research and Development, Abu Dhabi Food Control Authority, Abu Dhabi, UAE

# References

[1] Bernardi C. Nutritional Facts in Vinal (*Prosopis ruscifolia*) Pulp Including its Iron Availability [MS Thesis]. Santa Fe, Argentina: Faculty of Chemical Engineering: National University of Litoral; 2000

[2] USNAS (United States National Academy of Sciences). Firewood Crops: Shrub and Tree Species for Energy Production. Washington, DC: National Academy Press; 1980. pp. 150-151. DOI: https://doi.org/10.17226/21317

[3] Liu Y, Singh D, Nair MG. Pods of Khejri (*Prosopis cineraria*) consumed as a vegetable showed functional food properties. Journal of Functional Foods. 2012;**4**:116-121. DOI: 10.1016/j.jff.2011.08.006

[4] Al-Aboudi A, Afifi FU. Plants used for the treatment of diabetes in Jordan: A review of scientific evidence. Pharmaceutical Biology. 2010;**49**:221-239. DOI: 10.3109/13880209.2010.501802

[5] George C, Lochner A, Huisamen B. The efficacy of *Prosopis glandulosa* as antidiabetic treatment in rat models of diabetes and insulin resistance. Journal of Ethnopharmacology. 2011;**137**:298-304. DOI: https://doi.org/10.1016/j.jep.2011.05.023

[6] Garg A, Mittal SK. Review on *Prosopis cineraria*: A potential herb of Thar desert. Drug Invention Today. 2013;**5**:60-65. DOI: 10.1016/j.dit.2013.03.002

[7] Burkart A. Leguminosas. In: Carlos S. Flora de la Provincia de Buenos Aires. Buenos Aires: Colecc. Cientifica del INTA; 1967. Part 3. p. 394-647

[8] Burkart A. A monograph of the genus *Prosopis* (Leguminosae sub. Fam. Mimosoideae). Journal of the Arnold Arboretum. 1976;**57**(4):219-249, 450-525

[9] Leakey RRB, Last FT. Biology and potential of *Prosopis* species in and environments with particular reference to *P. cineraria*. Journal of Arid Environments. 1980;**3**:9-24

[10] Khan MAR. The Indigenous Trees of the United Arab Emirates: Dubai Municipality. Dubai, UAE; 1999. 150p

[11] Pasiecznik NM, Harris PJC, Smith SJ. Identifying Tropical *Prosopis* Species: A Field Guide. Association HDR. Coventry, UK: HDRA Publishing; 2004. 26 p

[12] Jongbloed M, Gary F, Benno B, Westren AR. The Comprehensive Guide to the Wild Flowers of the United Arab Emirates. Abu Dhabi, UAE: Environmental Research and Wildlife Development Agency; 2003. p. 576

[13] Silva MA. Taxonomy and distribution of the genus *Prosopis* L. In: Habit MA, Saavedra JC, editors. The Current State of Knowledge on Prosopis Juliflora. Rome, Italy: FAO; 1990. pp. 177-186

[14] Hafeez M. *Prosopis cineraria* (L.) Druce: Its Production and Management and Use, Pakistan. Bangkok, Thailand: FAO; 1991. p. 77

[15] Arshad M, Ashraf M, Arif N. Morphological variability of *Prosopis cineraria* (L.) Druce, from the Cholistan Desert, Pakistan. Genetic Resources and Crop Evolution. 2006; **53**(8):1589-1596. DOI: 10.1007/s10722-005-8563-5

[16] Lee SG, Felker P. Influence of water/heat stress on flowering and fruiting of mesquite (*Prosopis glandulosa* var. *glandulosa*). Journal of Arid Environmets. 1992;**23**:309-319

[17] Rani B, Singh U, Sharma R, Gupta A, Dhawan N, Sharma A, Maheshwari R. *Prosopis cineraria* (L.) druce: A desert tree to brace livelihood in Rajasthan. Asian Journal Pharmaceutical Research & Health Care. 2013;**5**(2):58-64

[18] Bhansali RR. Biology and multiplication of *Prosopis* species grown in the Thar Desert. In: Ramawat KG, editor. Desert Plants Biology and Biotechnology. Berlin Heidelberg, New York, USA: Springer; 2010. pp. 371-406. DOI: 10.1007/978-3-642-02550-1_18

[19] Khatri A, Rathore A, Patil UK. *Prosopis Cineraria*–A boon plant for desert. International Journal of Biomedical and Advanced Research. 2010;**1**(5):141-149. DOI: 10.7439/ijbar. v1i5.14

[20] Ramoliya PJ, Patel HM, Joshi JB, Pandey AN. Effect of salinization of soil on growth and nutrient accumulation in seedlings of *Prosopis cineraria*. Journal of Plant Nutrition. 2006;**29**:283-303. DOI: 10.1080/01904160500476806

[21] Khasgiwal PC, Mishra GG, Mithal BM. Studies on *Prosopis spicigera* gum. Part II. Emulsifying properties and HLB value. The Indian Journal of Pharmacy. 1970a;**32**:82

[22] Lemons J, Victor R, Schaffer D. Conserving Biodiversity in Arid Regions: Best Practices in Developing Nations. US: Springer; 2003. p. 497. DOI: 10.1007/978-1-4615-0375-0

[23] Ahmed S, Tasleem F, Azhar I. Samanea Saman and *Prosopis cineraria*: Traditional Use–Biological and Phytochemical Investigation. Saabrucken Deutschland, Germany: LAP LAMBERT Academic Publishing; 2013. 200 p. DOI: 10.13140/2.1.1175.6485

[24] Robertson S, Narayanan N, Deattu N, Ravi Nargis NR. Comparative anatomical features of *Prosopis cineraria* (L.) Druce and Prosopis juliflora. International Journal of Green Pharmacy. 2010;**4**(4):275-280. DOI: 10.22377/ijgp.v4i4.160

[25] Khasgiwal PC, Mithal BM. Studies on Prosopis spicigera gum. Part II: Emulsifying properties and HLB value. The Indian Journal of Pharmacy. 1970b:32-82

[26] Mann HS, Shankarnarayan KA. The role of *Prosopis cineraria* in an agropastoral system in western Rajasthan. In: Le Houerou HN, editor. Browse in Africa. Addis Ababa: International Livestock Center for Africa; 1980. p. 491

[27] Yadav RS, Yadav BL, Chhipa BR. Litter dynamics and soil properties under different tree species in a semi-arid region of Rajasthan, India. Agroforestry Systems. 2008;**73**:1-12. DOI: 10.1007/s10457-008-9106-9

[28] Gates PJ, Brown K. Acacia tortilis and *Prosopis cineraria*: Leguminous trees for arid areas. Outlook On Agriculture. 1988;**17**:61-64

[29] Sajwani A, Farooq SA, Bryant VM. Studies of bee foraging plants and analysis of pollen pellets from hives in Oman. Palynology. 2014;**38**:207-223. DOI: 10.1080/01916122.2013.871652

[30] Balogun AM, Fetuga BL. Chemical composition of some underexploited leguminous crop seeds in Nigeria. Journal of Agricultural and Food Chemistry. 1986;**34**(2):189-192. DOI: 10.1021/jf00068a008

[31] Jahreis G, Brese M, Leiterer M, Schäfer U, Böhm V, Jena. Legume flours: Nutritionally important sources of protein and dietary fiber. Ernaehrungs Umschau international. 2016:**2**:36-42. DOI: 0.4455/eu.2016.007

[32] Breene WM, Lin S, Hardman L, Orf J. Protein and oil content of soybeans from different geographic locations. Journal of the American Oil Chemists' Society. 1988;**65**(12): 1927-1931. DOI: https://doi.org/10.1007/BF02546009

[33] Gangal S, Sharma S, Rauf A. Fatty acid composition of *Prosopis cineraria* seeds. Chemistry of Natural Compounds. 2009;**45**(5):705-707. DOI: https://doi.org/10.1007/s10600-009-9425-8

[34] Joshi P, Nathawat NS, Chhipa BG, Hajare SN, Goyal M, Sahu MP, Singh G. Irradiation of sangari (*Prosopis cineraria*): Effect on composition and microbial counts during storage. Radiation Physics and Chemistry. 2011;**80**:1242-1246. DOI: https://doi.org/10.1016/j.radphyschem.2011.05.009

[35] Chandra J, Mali MC. Nutritional evaluation of top five fodder tree leaves of Mimosaceae family of arid region of Rajasthan. International Journal of Innovative Research and Review. 2014;**2**(1):14-16

[36] Ghazanfar S, Latif A, Mirza IH, Nadeem MA. Macro-minerals concentrations of major fooder tree leaves and shrubs of district chakwal, Pakistan. Pakistan Journal of Nutrition. 2011;**10**(5):480-484

[37] Felger RS. Ancient crops for the 21st century. In: Ritchle GA, editor. New Agricultural Crops. USA: Boulder: Colorado, Westview Press; 1979. pp. 5-20. DOI: https://doi.org/10.1017/S001447970001173X

[38] Felker P, Bandurski RS. Uses and potential uses of leguminous trees for minimal energy input agriculture. Economic Botany. 1979;**33**(2):172-184. DOI: https://doi.org/10.1007/BF02858286

[39] Becker R, Grosjean OK. A compositional study of pods of two varieties of mesquite (*P. glandulosa, P. velutina*). Journal of Agricultural and Food Chemistry. 1980;**28**(1):22-25. DOI: 10.1021/jf60227a024

[40] Orwa C, Mutua A, Kindt R, Jamnadass R, Simons A. *Prosopis cineraria*. Agroforestree Database: A Tree Reference and Selection Guide Version 4.0. Available from: http://www.worldagroforestry.org/af/treedb/ [Accessed: April 10, 2018]. 2009

[41] Panwar D, Pareek K, Bharti CS. Unripe pods of *Prosopis cineraria* used as a vegetable (sangri) in Shekhawati region. International Journal of Scientific & Engineering Research. 2014;**5**(2):892-895

[42] Cattaneo F, Costamagna MS, Zampini IC, Sayago J, Alberto MR, Chamorro V, Pazos A, Thomas-Valdés S, Schmeda-Hirschmann G, Isla MI. Flour from *Prosopis* alba cotyledons: A natural source of nutrient and bioactive phytochemicals. Food Chemistry. 2016;**208**:89-96. DOI: https://doi.org/10.1016/j.foodchem.2016.03.115

[43] Saura-Calixto F. New food products from *Prosopis* fruits in Latin America: A base for the extension of the culture and prevention of desertification in arid zones. pp. 265-268. In: Summary reports of European Commission supported STD-3 projects (1995-1997);1999. CTA. www.agricta.org/pubs/std/vol2/pdf/341.pdf

[44] Felker P, Takeoka G, Dao L. Pod Mesocarp flour of north and SouthAmerican species of leguminous TreeProsopis (mesquite): Composition and food applications. Food Reviews International. 2013;**29**:49-66. DOI: 10.1080/87559129.2012.692139

[45] Choge SK, Harvey M, Chesang S, Pasiecznik NM. Cooking with Prosopis flour. In: Recipes Tried and Tested in Baringo District, Kenya. Coventry, UK: KEFRI Nairobi, Kenya and HDRA; 2006. pp. 1-6

[46] Schmeda-Hirschmann G, Razmilic I, Gutierrez MI, Loyola JI. Proximate composition and biological activity of food plants gathered by Chilean Amerindians. Economic Botany. 1999;**53**:177-187. DOI: https://doi.org/10.1007/BF02866496

[47] Astudillo L, Schmeda-Hirschmann G, Herrera JP, Cortés M. Proximate composition and biological activity of Chilean *Prosopis* species. Journal of the Science of Food and Agriculture. 2000;**80**:567-573. DOI: 10.1002/(SICI)1097-0010(200004)80:5<567::AID-JSFA563>3.0.CO;2-Y

[48] Cardozo ML, Ordóñez RM, Zampini IC, Cuello AS, Di Benedetto G, Isla MI. Evaluation of antioxidant capacity, genotoxicity and polyphenol content of non-conventional foods: *Prosopis* flour. Food Research International. 2010;**43**(5):1505-1510. DOI: https://doi.org/10.1016/j.foodres.2010.04.004

[49] Cattaneo F, Sayago J, Alberto MR, Zampini IC, Ordoñez RM, Chamorro V, Pazos A, Isla MI. Anti-inflammatory and antioxidant activities, functional properties and mutagenicity studies of protein and protein hydrolysate obtained from Prosopis alba seed flour. Food Chemistry. 2014;**161**:391-399. DOI: https://doi.org/10.1016/j.foodchem.2014.04.003

[50] Pérez MJ, Cuello AS, Zampini IC, Ordóñez RM, Alberto MR, Quispe C, Isla MI. Polyphenolic compounds and anthocyanin content of *Prosopis nigra* and *Prosopis alba* pods flour and their antioxidant and anti-inflammatory capacity. Food Research International. 2014;**64**:762-771. DOI: https://doi.org/10.1016/j.foodres.2014.08.013

[51] Tull D, Larke J, Teaque J, Rippe S. Edible and Plants in Edible and Useful Plants of Texas and the Southwest : Texas, New Mexico, and Arizona. 3ed ed. University Texas Press, Austin; 2013. 91P

[52] Busch VM, Delgado JF, Santagapita PR, Wagner JR, Buera MP. Rheological characterization of vinal gum, a galactomannan extracted from *Prosopis* ruscifolia seeds. Food Hydrocolloids. 2018;**74**:333-341. DOI: https://doi.org/10.1016/j.foodhyd.2017.08.010

[53] El Hag MG, Shargi KM, Eid AA. The nutrient composition of animal feeds available in the Sultanate of Oman. Ag+.5riculture and Fisheries Research Bulletin, Ministry of Agriculture and Fisheries, Sultanate of Oman. 2000;**1**:1-14

[54] Mahgoub O, Kadim I, Al-Ajmi D, Al-Saqry NM, Al-Abri AS, Richie AR, Annamalai K, Forsberg NE. The effects of replacing Rhodes grass (Chloris gayana) hay with Ghaf (*Prosopis cineraria*) pods on the performance of Omani native sheep. Tropical Animal Health and Production. 2004;**36**:281-294. DOI: 10.1023/B:TROP.0000016829.63985.2c

[55]  Bohra HC. Nutrient utilization of *Prosopis cineraria* (Khejri) leaves by desert sheep and goat. Annals of Arid Zone. 1980;**19**:73-81

[56]  Prasad R, Kumar R, Vaithiyanathan S, Patnayak BC. Effect of polyethylene glycol 4000 treatment upon nutrient utilization from Khejri (*Prosopis cineraria*) leaves in sheep. The Indian Journal of Animal Science. 1997;**67**(8):712-715

[57]  Bhatta R, Shinde AK, Vaithiyanathan S, Sankhyan SK, Verma DL. Effect of polyethylene glycol-6000 on nutrient intake, digestion and growth of kids browsing *Prosopis cineraria*. Animal Feed Science Technology. 2002;**101**(1):45-54. DOI: 10.1016/S0377-8401(02)00180-3

[58]  Vaithiyanathan S, Bhatta R, Mishra AS, Prasad R, Verma DL, Singh NP. Effect of feeding graded levels of *Prosopis cineraria* leaves on rumen ciliate protozoa, nitrogen balance and microbial protein supply in lambs and kids. Animal Feed Science Technology. 2007;**133**:177-191. DOI: 10.1016/j.anifeedsci.2006.04.003

[59]  Bhatta R, Vaithiyanathan S, Singh NP, Verma DL. Effect of feeding complete diets containing graded levels of *Prosopis cineraria* leaves on feed intake, nutrient utilization and rumen fermentation in lambs and kids. Small Ruminant Research. 2007;**67**(1):75-83. DOI: 10.1016/j.smallrumres.2005.09.027

[60]  Malik S, Mann S, Gupta D, Gupta RK. Nutraceutical properties of *Prosopis cineraria* (L.) Druce pods: A component of "Panchkuta". Journal Of Pharmacognosy and Phytochemistry. 2013;**2**(2):66-73

[61]  Rastogi RP, Mehrotra BN, editors. Compendium of Indian Medicinal Plants. Publication and Information Directorate: CSIR, New Delhi; 1993c. Vol III): 531 p

[62]  Salar RK, Dhall A. Antimicrobial and free radical scavenging activity of extracts of some Indian medicinal plants. Journal of Medicinal Plants Research. 2010;**4**:2313-2320. DOI: 10.5897/JMPR10.155

[63]  Preeti K, Avatar SR, Mala A. Pharmacology, phytochemistry and therapeutic application of *Prosopis cineraria* linn: A review. Journal of Plant Sciences. 2015;**3**(1-1):33-39. DOI: 10.11648/j.jps.s.2015030101.15

[64]  Sharma N, Garg V, Paul A. Antihyperglycemic, antihyperlipidemic and antioxidative potential of *Prosopis cineraria* bark. Indian Journal of Clinical Biochemistry. 2010;**25**(2):193-200. DOI: https://doi.org/10.1007/s12291-010-0035-9

[65]  Pareek AK, Garg S, Kuma M. *Prosopis cineraria*: A gift of nature for pharmacy. International Journal of Pharmaceutical Sciences and Research. 2015;**6**:958-964

[66]  Purohit A, Ram H. Hypolipidemic and antiatherosclerotic effects of *Prosopis cineraria* bark extract in experimentally induced hyperlipidemic rabbits. Asian Journal of Pharmaceutical and Clinical Research. 2012;**5**(3):106-109

[67]  Robertson S, Narayanan N, Kapoor BR. Antitumour activity of *Prosopis cineraria* (L.) Druce against Ehrlich ascites carcinomainduced mice. Natural Product Research. 2011; **25**(8):857-862. DOI: https://doi.org/10.1080/14786419.2010.536159

[68] Bahorun T, Necgheen VS, Aruoma OI. Botanical drugs, nutraceuticals and functional foods: The context of Africa. In: Bagchi D, editor. Nutraceutical and Functional Food Regulations in the United States and around the World. New York, NY, USA: Academic Press; 2008. pp. 341-348

[69] Simpson BB, Nedd JL, Moldenke AR. *Prosopis* flowers as a resource. In: Mesquite, its Biology in two Desert Shrub Ecosystems. Stroudsburg, Pennsylvania, USA: Dowden, Hutchinson and Ross; 1977. pp. 84-107

[70] NAS. Firewood Crops: Shrub and Tree Species for Energy Production. Washington DC, USA: National Academy of Sciences; 1980. DOI: https://doi.org/10.17226/19480

[71] Sharma BM. Chemical analysis of some desert tree. In: proceedings of the Symposium on Recent Advances in Tropical Ecology. Varansi: India; 1968. p. 248-251

[72] Malik A, Kalidhar SB. Phytochemical examination of *Prosopis cineraria* L. Druce) leaves. Indian Journal of Pharmaceutical Sciences. 2007;**69**(4):576-578. DOI: 10.4103/0250-474X. 36950

[73] Bahuguna U, Shuklam RN. A study on investigation of the chemical constituents & milled wood lignin analysis of *Lantana camara* & *Prosopis* chinnsis. International Journal of Applied Biology and Pharmaceutical Technology. 2010;**1**(3):830-835

[74] Dangar RD, Verma PD, Dangar RR, Patel JB, Patel KN. Phytopharmacological potential of Prosopis spicigera Linn. American Journal of PharmaTech Research. 2012;**2**(2):46-49

[75] Singh S, Naresh V, Kr SS. Pharmacognostic studies on the leaves of *Prosopis cineraria* (L.) Druce. growing in South Haryana. India Journal of Pharmacognosy and Phytochemistry. 2013;**2**(1):320-325

[76] Pareek AK, Garg S, Kumar M. Mal yadav S. *Prosopis Cineraria*: A gift of nature for pharmacy. International Journal of Pharma Sciences and Research. 2015;**6**(6):968-964

[77] Chopra RN, Chopra IC, Handa KL, Kapur LD. In: Chopra RN, Chopra IC, editors. Indigenous Drugs of India. Calcutta: UN. Dhur and Sons Pvt. Ltd; 1958. p. 521

[78] Chopra RN, Nayar SL, Chopra ICI. Glossary of Indian Medicinal Plants. New Delhi: Council for Scientific and Industrial Research; 1956. 204 p

[79] Nadkarni AK. Indian Materia Medica. Bombay: Popular Book Depot; 1954. p. 1011

[80] Malik A, Kalidar SB. Phytochemical examination of Prosopis cineraria L. (Druce) Leaves. Indian J. Pharmaceutical Sciences. 2007;**7-8**:576-578. DOI: 10.4103/0250-474X.36950

[81] Shalini. Vedic Leguminous Plants: Medicinal and Microbiological Study. Classical Pub.; 1997;**1**:57-58

[82] Persia FA, Rinaldini E, Hapon MB, Gamarra-Luques C. Overview of Genus Prosopis Toxicity Reports and its Beneficial Biomedical Properties. Journal of Clinical Toxicology. 2016;**6**:326-333. DOI: 10.4172/2161-0495.1000326

[83] Velmurugan V, Arunachalam G, Ravichandran V. Antibacterial activity of stem bark of *Prosopis cineraria* (Linn.) Druce. Archives of Applied Science Research. 2010;**2**(4):147-150

[84] Rastogi RP, Mehrotra BN. Compendium of Indian Medicinal Plants. Publication and Information Directorate. Vol I. New Delhi: CSIR; 1993a. 561 p

[85] Rastogi RP, Mehrotra BN. Compendium of Indian Medicinal Plants. Publication and Information Directorate. Vol II. New Delhi: CSIR; 1993b. 561 p

[86] Rastogi RP, Mehrotra BN. Compendium of Indian Medicinal Plants. Publication and Information Directorate. Vol IV. New Delhi: CSIR; 1993d. 597 p

[87] Khandelwa P, Sharm RA, Agarwal M. Phytochemical analyses of various parts of *Prosopis cineraria*. International Journal of Pharmacy and Chemistry. 2016;**2**(1):6-9